Digital Transformation of Industrial Processes

Digital Transformation of Industrial Processes

Advanced Features, Digital Twin, Cloud Computing, Intelligent Control

Khaled Benfriha

iSTE

WILEY

First published 2026 in Great Britain and the United States by ISTE Ltd and John Wiley & Sons, Inc.

ISTE Ltd
27-37 St George's Road
London SW19 4EU
UK

www.iste.co.uk

John Wiley & Sons, Inc.
111 River Street
Hoboken, NJ 07030
USA

www.wiley.com

Library of Congress Control Number: 2026941475

British Library Cataloguing-in-Publication Data
A CIP record for this book is available from the British Library
ISBN 978-1-83669-050-4

The manufacturer's authorized representative according to the EU General Product Safety Regulation is Wiley-VCH GmbH, Boschstr. 12, 69469 Weinheim, Germany, e-mail: Product_Safety@wiley.com.

Contents

Preface

In a perpetually evolving world, the digitalization of production systems has become an unavoidable necessity. This dynamic is fueled by internal company factors, such as the integration of new functionalities and the pursuit of operational performance, as well as by external requirements emanating from the ecosystem, such as the rigorous traceability of components during manufacturing or real-time monitoring reports available on shared platforms. Industry 4.0, with its promises of flexibility, autonomy and optimization, is redefining the contours of modern manufacturing.

This book aims to explore the multiple dimensions of this transformation in the initial chapters, while highlighting the key concepts that structure the digital transition of production systems in the following chapters. By drawing on proven transition methods used within the Industry 4.0 platform of the École Nationale Supérieure d'Arts et Métiers (ENSAM), we will provide enriching feedback and approaches to better frame the transition phases. We will also address the complexity and risks to consider during the preparatory stages.

We will analyze how solutions such as intelligent computing and the digital twin can provide adapted responses to contemporary challenges. Furthermore, we will examine the role of dynamic control systems, which facilitate real-time scheduling of operations, thereby ensuring agility and responsiveness within production chains. Additionally, we will propose design architectures that integrate the communicational, functional and organic dimensions of digitalized industrial systems.

I wish to express my profound gratitude to the doctoral students who contributed to this research; their commitment and ideas were essential. I extend particular thanks to Patrick Le Men, who directed several specialized apprenticeship training centers for engineering programs. His help and advice were invaluable throughout this project, enriching this work and deepening the knowledge presented here. Finally, I thank all colleagues as well as the institutional and industrial partners who made this work possible and relevant.

This book is intended for professionals, researchers and students wishing to understand the implications of this digital transformation. It is designed as a practical guide based on tested approaches, offering insightful perspectives on methods for transitioning to digitalized industrial systems. We hope that these reflections will contribute to a deep understanding of the challenges of the digital transition and will encourage innovative initiatives within industry.

Welcome to this fascinating and complex universe, where communicating machines and artificial intelligence (AI) unite to shape the industrial production systems of tomorrow.

June 2026

Introduction

The principle of industrial digitalization was first presented in 2011 at the Hannover Industrial Technology Fair. Also called Industry 4.0, the Industry of the Future, or the Smart Factory, it corresponds to the Fourth Industrial Revolution in the continuity of previous revolutions. The history of industrial progress has previously experienced three revolutions. The first appeared at the end of the 17th century in the United Kingdom, the second at the end of the 19th century in the United States and Germany, and the third at the end of the 20th century in the United States and Japan.

Generally, the factors that initiate an industrial revolution are always the same: technological inventions or innovations, a favorable market and investors willing to take risks. Steam, iron and coal gave rise to the First Industrial Revolution, which propelled England to the rank of an economic power in the 18th century. The second, which occurred between 1890 and 1910, emerged in Germany and the United States. It was founded on new energy sources such as electricity and oil, and on heavy industry. The third, which took place between 1970 and 2000, began in the United States and Japan. Initiated by multiple technological innovations in information, communication, nuclear power and the transportation industry, it thus paved the way for globalization.

Since then, the world has experienced a fourth revolution, which ushered in the hyper-connected factory through new information and communication technologies (ICT), thereby offering new functionalities better adapted to the new ecosystem. Climate disruptions and their consequences are prompting some experts to envision a new consumption model, one that is more sober and sustainable. Concurrently, reflections are emerging on the place of humans in this global transformation. These questions open the way to a potential Fifth Industrial Revolution, one that would extend far beyond mere technological advances.

All industrial revolutions have brought about a new economic dynamic due to productivity gains, thereby offering new services or making products accessible to a large number of consumers. They also entail societal and human evolutions. Despite the improvement in working conditions, the working population has often been led to change professions or adapt, to migrate to employment hubs, to accept strict task control, or even to adopt inappropriate postures that can impact health.

Humans have not always been at the center of industrial transformations. The major economic models upon which industrial investment projects are built prioritize financial performance. However, the consequences of the Second and Third Industrial Revolutions on the environment are now indisputable. Soil pollution, global warming and fine particulate matter have awakened a collective consciousness regarding the necessity of adopting a new consumption model, one that is sustainable, decarbonized, equitable and centered on human social well-being. Due to the new functionalities it offers, the Fourth Industrial Revolution partially fits into this framework. While smart factories can reduce energy footprints, recycle waste on-site, limit raw material usage, etc., their development also raises questions, some of which are strategic for companies, such as cybersecurity or dependence on digital platforms that provide tools for the digitalization and control of production systems.

It is undeniable that previous industrial revolutions propelled humanity into progress beyond all measure. Who could have imagined a mobile scientific laboratory exploring the planet Mars thanks to embedded artificial intelligence (AI)? But at the same time, they generated north–south inequalities and irreversible climatic and environmental consequences. The challenge now consists of inscribing the Fourth, or by extension the Fifth, Industrial Revolution into a different framework, one that places on the same scale economic performance, human well-being and the preservation of the planet.

On the one hand, it is certain that companies must integrate these upheavals now to avoid suffering the consequences in the future. Regardless of their size, companies play a major role in society: they do not merely provide products or services but are also living environments where employees invest themselves to create added value, and their activity must endure and prosper. The digitalization of industrial processes can constitute a partial response to achieving objectives such as decarbonization, personalization, local production, recyclability, sustainability, maintainability, sharing and even adaptation to telework, within the framework of a new, transparent socioeconomic model.

We are living through a pivotal period in which humanity is becoming aware of the necessity to change its behaviors, particularly regarding the consumption of products and services. Furthermore, new technologies offer promising solutions to meet these challenges. It now falls to political decision-makers to communicate that this necessary transition generates costs.

In this context, this book explores the transition of companies toward the digitalization of their production systems. It is aimed at the general public as well as students and professionals wishing to understand the challenges of this transformation, particularly in terms of methodology, opportunities, employment, risks and technological integration. The initial chapters offer an overview of the fundamentals of Industry 4.0 and describe the main concepts that interact to form a digitalized production line. Finally, a detailed presentation of the Industry 4.0 platform is provided, in which solutions were experimented with by researchers from the École Nationale Supérieure d'Arts et Métiers de Paris (ENSAM). The following chapters present concrete experiments implementing four functionalities adapted to existing installations.

Given the breadth of the subject of the factory of the future, priorities have been established to focus on transition, while keeping in mind that certain areas will require more in-depth exploration in the future. This book aims to provide a foundation of essential knowledge as well as concrete feedback, enabling a proper understanding of difficulties while benefiting from ideas and advice for the deployment phase. Readers will observe that the transition to digitalized industrial systems is not linear and requires additional engineering work, the costs of which are often unforeseen.

The first chapter explores the complexity of the digital age and its impact on employment and training, highlighting how the digital transition is redefining economic and professional paradigms. It analyzes current research, proposes global strategies and examines emerging economic models, shedding light on the challenges and opportunities of the digital revolution. The second chapter presents a synthesis of the new functionalities offered by Industry 4.0, grouped into three categories: "preparation", "supervision" and "cross-functional".

The third chapter presents the main technological concepts supporting a digitalized industrial system. These concepts, which are interdependent and essential for a successful transition, include the notions of data and information, cyber-physical manufacturing systems, the industrial internet of things (IIoT), cyber-physical industrial systems, dynamic control and intelligent computing. The fourth chapter addresses the notions of modeling and architecture, including general concepts, a reference model, an architecture given as an example and a hybrid transition model. The notion of hybrid refers to an intermediate model between a less flexible, pyramid-type version and a fully distributed version that is less suited to the needs of companies with conventional production tools. Thus, the advantage of this intermediate model is that it facilitates the integration of existing industrial installations and equipment.

The fifth chapter is dedicated to the Industry 4.0 platform. In 2015, ENSAM launched an ambitious project aimed at developing a full-scale production line. The objective of this project is to conduct research to acquire new knowledge, to validate and evaluate the effectiveness of proposed solutions and to strengthen engineering training programs with skills adapted to the requirements of Industry 4.0. At the end of this chapter, a study is presented on the integration of a 3D scanner as a cyber-physical industrial system dedicated to quality control within the Industry 4.0 platform.

The sixth chapter explores advanced control systems, which are essential for achieving performance objectives. These systems use technologies such as artificial intelligence (AI), advanced data analytics and real-time connectivity to improve decision-making, optimize flow, reduce downtime and minimize energy consumption. The seventh chapter examines the principles of a digital twin from three perspectives: maturity levels, categorization and execution mode. It then details an approach for designing a digital twin, the experimentation of which was carried out on the Industry 4.0 platform, highlighting the technological challenges and the importance of multidisciplinary skills.

The eighth chapter is dedicated to cloud manufacturing, a promising technology for the digitalized industry. It is structured into four sections: first, a general classification of the different types of clouds, followed by an innovative methodology and framework for integrating and adapting cloud resources to the needs of digitalized production systems. Finally, the ninth chapter addresses the "optimization" functionality in the context of Industry 4.0. Indeed, the massive data collected by cyber-physical systems and the IIoT network, combined with the powerful computational resources of cloud computing, open new optimization perspectives. To this end, an experimental methodological approach was proposed and implemented on a real case of cutting condition optimization. An algorithmic pipeline, which was designed using cloud tools, makes it possible to determine the cutting parameters that minimize energy consumption, vibrational disturbances and operation time. The results obtained show that a structured integration of different Industry 4.0 technologies is promising, but it also reveals an intrinsic complexity, articularly in the context of multi-objective dynamic optimization.

1

Impacts and Challenges of Industry 4.0

Industry 4.0, marked by the integration of advanced digital technologies into production systems, is profoundly transforming the manufacturing sector and beyond. It offers considerable opportunities to improve performance and profitability. However, it also poses significant challenges in terms of cybersecurity, equity, skills and sustainability. In this context, a multitude of multidisciplinary research efforts are striving to provide avenues for reflection on methodological, technological and algorithmic aspects, thereby contributing to making the transition more structured.

The integration of advanced technologies is changing not only the way we work but also the content of jobs. A balanced approach, one that recognizes both the advantages and the challenges, is crucial for a successful transition and for ensuring that the benefits of Industry 4.0 profit society as a whole. However, concerns are emerging regarding the quality of jobs created, working conditions and wages, especially for less specialized skills, which can accentuate socioeconomic inequalities. Similarly, the adoption of cutting-edge technologies could induce a form of discreet surveillance in the workplace, thereby affecting employee well-being. Although training or the development of new skills are frequently proposed as solutions to preserve employment, in reality, no guarantee is given.

Furthermore, the digital transformation of production systems is establishing itself as an essential strategy for companies wishing to prepare advantageously for the future. Its success rests on relevant choices in terms of targeted functionalities, interaction with the ecosystem, employee involvement and economic models, which offer new opportunities to further enhance the value of digitalized production systems. But the path is far from easy, particularly due to technological evolutions, the lack of experts, structural constraints, cyber risks, dependence on technology suppliers and other parameters that generate additional costs. This underscores the need for a comprehensive and sustainable strategy, including a strategic vision and reasonable orientations.

This chapter examines the complexity of the digital age and its impact on employment and training, highlighting how the industrial digital transition is redefining economic and professional paradigms. It explores current research accompanying this transformation and proposes a comprehensive approach for undertaking such projects. By analyzing emerging economic models and the central role of data, the chapter aims to offer a general understanding of the challenges and opportunities of the digital revolution, providing perspectives on how technology is transforming the world at the micro- and macroeconomic levels.

1.1. Employment and training

Industry 4.0 has a significant and diversified impact on society, directly affecting the world of work, the dynamics of socio-professional relations and learning methods, particularly through the acquisition of digital skills adapted to increasingly complex technologies. This industrial revolution responds to customers who are more attentive to responsible consumption practices, such as supporting short supply chains, reducing ecological footprints and preferring products that are personalized, shareable, repairable and recyclable. Furthermore, this industrial era raises crucial questions concerning the position of humans in terms of interactions, employment, tasks, etc.

The autonomy of industrial systems, characterized by the increasing integration of advanced technologies such as artificial intelligence (AI) and mobile robotics, is on the verge of revolutionizing the world of work as we know it. This evolution, although promising progress and efficiency, raises important questions regarding its impact on the sustainability of certain jobs. In particular, positions characterized by repetitive tasks with low variability, which represent a significant portion of the labor market in many industrial sectors, are likely to be the most affected.

In the manufacturing sector, jobs involving repetitive manual tasks such as assembly line work and packaging are greatly threatened by autonomous and robotic systems. Furthermore, increased efficiency and process optimization can also lead to a smaller number of employees being necessary to achieve the same production levels. This is already perceptible in the transportation and agriculture sectors.

The advent of autonomous vehicles could also impact transportation jobs, such as those of truck, taxi and delivery drivers. If autonomous vehicles are widely adopted, this could lead to a decrease in employment opportunities for drivers, given that driving tasks could be performed more economically and safely by embedded and intelligent control systems.

In agriculture, the rise of digitalization has transformed these machines into intelligent systems capable of making decisions based on real-time data. For example, tractors equipped with sensors and actuators can control operations day and night to maximize the efficiency of plowing, planting and harvesting. AI plays a crucial role in analyzing the data collected by digitalized agricultural machines, offering insights to optimize yields and reduce both the use of resources (water, fertilizers and pesticides) and the need for personnel.

However, this transition toward greater autonomy of industrial systems also opens the door to the emergence of new jobs and the need for new skills. The programming, maintenance, monitoring and improvement of these autonomous systems require specialized expertise. Furthermore, the ability to work in synergy with these technologies, to understand and interpret their operations, and to intervene in the event of anomalies or necessary adjustments becomes paramount.

It is difficult to obtain precise forecasts on the impact of Industry 4.0 on employment. However, several reports attempt to quantify the effects of advanced technology integration on employment in a general manner, highlighting the duality of Industry 4.0's impact on employment. The "The Future of Jobs Report" from the World Economic Forum (WEF 2023) reveals, on the one hand, that 75% of companies plan to adopt Big data, cloud computing and AI by 2027. Furthermore, the study suggests that 69 million jobs will be created, resulting from an adaptation and division of labor between humans, machines and algorithms. At the same time, 83 million jobs will be destroyed, leading to a contraction of global labor markets by 14 million jobs at the current pace of evolution, although this figure is subject to a high degree of uncertainty.

With the advent of Industry 4.0, the need for advanced technical skills is increasing. Workers must now understand and interact with cutting-edge technologies, which requires continuous training to remain current. Skills in programming, data analytics, cybersecurity, robotics and information systems management are becoming essential. Furthermore, as technologies evolve rapidly, workers must constantly acquire new knowledge to stay relevant. This implies a transition toward flexible learning models, including online learning, microcredentials and blended learning, which combine traditional and digital methods.

Thanks to AI and big data, it is possible to create learning pathways adapted to the needs and pace of each learner. This can help fill skills gaps more effectively and encourage more active participation. Other tools, such as virtual reality (VR) and augmented reality (AR), offer immersive learning opportunities. These technologies can simulate real work environments, allowing learners to practice in a risk-free setting without disrupting production. This is particularly useful for training on expensive or dangerous equipment. Not to mention online collaboration platforms and communication tools that enable knowledge exchange and social learning, thereby strengthening the collective capacity to solve complex problems. This new dynamic brings its share of challenges, particularly in terms of accessibility, data protection and curriculum updating.

Digitalized production systems, marked by advanced technological integration and sophisticated control systems, undeniably lead to a reduction in the number of operators, directly or indirectly, on production lines. This underscores the importance of close collaboration between public policies, educational institutions and companies themselves in order to rethink education and vocational training to make them more suitable. Anticipating these changes amounts to minimizing disruptions and maximizing opportunities in order to make this industrial revolution socially acceptable.

Indeed, the rapidity of technological change induced by Industry 4.0 can make newly acquired skills quickly obsolete, forcing workers to retrain or change careers, often without guarantee of success, which could, moreover, accentuate socioeconomic inequalities. Concurrently, the quality of jobs generated raises concerns, particularly regarding job security, working conditions and remuneration, with a tendency toward the devaluation of positions requiring less specialized skills. Furthermore, the widespread adoption of intelligent technologies can lead to indirect surveillance at work, eroding employee autonomy and potentially affecting their well-being.

The equation linking Industry 4.0 and employment presents a complex challenge for scientists, economists and policymakers. The discourses surrounding this transition are indeed contradictory. Some observers are optimistic, emphasizing that the digitalization of industrial processes can generate new jobs. According to this perspective, Industry 4.0 could be a source of unprecedented opportunities for those who possess the appropriate skills. In contrast, a more pessimistic current emphasizes the risks of job destruction. Skeptics point to the capacity of Industry 4.0 technologies to automate tasks previously performed by humans, exacerbating problems of unemployment and inequalities.

The complexity of the situation is such that it is currently difficult to establish a proportionality between the jobs that will be generated and those that will disappear due to Industry 4.0. Studies on this subject present widely varying estimates, due to numerous factors, including the public policies implemented to support this transition, the adaptability of the workforce, as well as the speed at which these technologies will be adopted and adapted in different industrial and geographic contexts. The manner in which different societies meet this challenge will largely determine whether Industry 4.0 will be a source of opportunities or divisions.

1.2. The industrial digital transition

The industrial digital transition refers to the process by which manufacturing and industrial companies integrate advanced digital technologies into all aspects of their operations. This transformation aims to improve the efficiency, productivity and flexibility of production processes, while adapting to changes in customer performance requirements and societal expectations. The industrial digital transition is structured around three main pillars. The first concerns the evolution of industrial system architecture, marked by a fundamental change: the shift from a traditional pyramid architecture to a distributed architecture. The second pillar is based on technological integration, particularly the transformation of production tools into interconnected "cyber-physical" entities. The third pillar addresses organizational aspects and dynamic control of operations.

Historically, the organization of production systems followed a rigid hierarchical structure, often represented in the form of a pyramid divided into several levels: at the top, production planning and management, or enterprise resource planning (ERP); followed by production control, or manufacturing execution system (MES); process control, or supervisory control and data acquisition (SCADA); and programmable logic controllers (PLCs); and at the bottom, the sensors and actuators that interact directly with the manufacturing process. With Industry 4.0, this pyramid architecture evolves toward a model that is more distributed, flexible, integrated and data-oriented. This new architecture is characterized by the decentralization of decisions and processes.

Intelligent control systems have the capacity to make decisions in real time, at the local level, without requiring centralized intervention. This makes production processes more responsive and flexible, allowing for better adaptation to changes and unforeseen events. Furthermore, interconnectivity between machines, systems and data across the entire enterprise and beyond (with partners, suppliers and customers) is another key characteristic. The traditional architecture, with its various compartmentalized levels, is giving way to open and fluid data exchange networks, promoting greater flexibility and optimization.

The data collected by sensors and the Internet of things (IoT) devices throughout the factory are no longer merely transmitted up the pyramid for analysis and action. Instead, thanks to AI and machine learning (ML), data can be analyzed locally to guide processes in real time or readjust local manufacturing parameters. To this end, edge computing is a technology dedicated to data processing at the network edge, close to the data source, to reduce latency and accelerate decision-making. This combination breaks the traditional hierarchy and enables an efficient distribution of computational resources.

In summary, Industry 4.0 redefines the architecture of industrial systems into a much more fluid, responsive and adaptive structure. This evolution toward horizontal and vertical integration, supported by advanced technologies, paves the way for smart factories where flexibility, predictive capability and portability are at the forefront, thus revolutionizing the manner in which industries operate and create value.

The integration of new technological components into existing production systems constitutes a major challenge for companies wishing to remain competitive. This process is complex and multifaceted, involving technical, organizational and human challenges. Existing production systems were often designed and implemented at a time when currently available technologies did not exist or were in their infancy. Consequently, these systems may be based on architectures, standards and protocols different from those of new technologies. Integration therefore requires technical compatibility, which may involve updating or modifying existing infrastructure, as well as developing interfaces to enable fluid communication between old and new systems.

The integration of new technologies, which generally relies on a set of specifications expressing the desired functionalities, involves the transformation of production machines and equipment into interconnected "cyber-physical" entities. Similarly, it often requires a revision of production processes, work practices and control models. Companies must adopt a holistic approach, involving reliable partners, to ensure that technical changes produce the expected effects. This may necessitate a restructuring of teams, a redefinition of roles and responsibilities, while being attentive to the uncertainties and complexity that are inherent to the integration and appropriation of new technologies.

The human aspect is often the most sensitive factor to manage in the integration process. New disruptive technologies can generate resistance within teams, due to fear of the unknown, concerns about the need to having new skills or even the legitimate and real fear of losing our job. Training and employee support are therefore crucial to ensure a successful transition. It is important to communicate clearly about the reasons for integrating new technologies, the expected benefits and the support plans provided for employees.

The integration of new technological components into existing production systems is therefore a complex and time-consuming process that extends far beyond purely technical considerations. It requires an integrated approach, taking into account multi-scale interoperability aspects, to overcome inherent challenges and fully benefit from new functionalities. Furthermore, the adoption of these technologies is crucial for companies seeking to remain competitive within interconnected ecosystems. In this environment, companies no longer operate in silos but integrate into a broader network, responding to a new approach to production and added value creation. This model highlights how vital it is for modern companies to evolve jointly by sharing their resources and data.

Moreover, Industry 4.0 is revolutionizing the control of industrial operations by introducing the concept of dynamic and autonomous control. The latter is based on the capacity to continuously adapt and optimize production processes in response to demand variations, unforeseen events, priorities or changes in the supply chain. Dynamic control of production operations presents several major advantages and innovations. Thanks to ubiquitous connectivity and real-time data exchange between machines, systems and actors, companies can react more quickly to demand fluctuations or other changes in the market environment. Thus, dynamic control makes it possible to adjust production plans, minimize downtime and optimize resource utilization.

Furthermore, the exploitation of big data and ML enables decision-makers to access valuable insights into operations, maintenance, quality and supply. The integration of AI offers relevant capabilities, ranging from equipment failure prediction to production flow optimization. However, the implementation of these concepts is complex due to the large number of parameters and variables to consider, making the analytical modeling of production systems and their resolution difficult. Moreover, this requires significant computational resources, particularly if we wish to activate multi-objective optimization algorithms dedicated to dynamic scheduling of operations, such as recalculating robot movement that maximizes speed while minimizing energy consumption without hindering flow dynamics – which can represent a major scientific challenge.

Dynamic and autonomous control is a prerequisite for flexible production and mass customization, enabling companies to produce goods adapted to individualized specifications without compromising efficiency or significantly increasing costs. This is made possible by autonomous and interconnected production systems that can easily be reconfigured for different products or variants. Similarly, dynamic control of operations encourages closer collaboration not only within the company but also with suppliers, partners and customers. Digital platforms and connected ecosystems facilitate information exchange and activity coordination along the value chain, thereby improving transparency and the capacity to respond coherently to market requirements.

In recent years, the industrial landscape has undergone a remarkable transformation, with a growing number of companies embarking on a strategic shift toward the digitalization of their production, despite certain resistance to change within their organizations. This evolution is not merely a trend but rather an adaptive response to consumer expectations and the demands of an evolving market. However, the transition to digitalized production also presents challenges. It requires significant investments in terms of technologies and training. Furthermore, it raises questions of cybersecurity and dependence on technology suppliers.

The economic dimension is often misleading, as certain costs are invisible at the outset but necessary if we wish to achieve the objectives. Indeed, it is not enough to acquire a manufacturing execution system (MES) platform to benefit from intelligent digital control of production systems; we must also first digitize the production lines and their functions to facilitate their integration into the platform. This work is entrusted to experts in various disciplines, such as robotics, industrial computing, automation, communication networks and mechatronics, while collaborating with different suppliers of production equipment or technological solutions. Additional engineering studies are often necessary and carried out by external service providers, causing additional costs and delays.

Despite these challenges, the impetus toward production digitalization has become a strategic necessity for companies seeking to position themselves advantageously for the future. In summary, the digital transition of manufacturing and industrial companies through the integration of advanced technologies is a progressive, risky and costly undertaking. Its success lies in the proper definition of desired functionalities, ecosystem requirements, employee engagement and the necessary technological minimum.

1.3. Scientific research

Research on Industry 4.0 is highly active and covers a wide range of domains, reflecting the breadth and depth of the changes that the Fourth Industrial Revolution brings. Research efforts focus not only on the development and improvement of underlying technologies but also on exploring their applications, assessing their socioeconomic impact and formulating frameworks for their effective integration. The scientific challenges associated with Industry 4.0, which arise directly from technological advances and their implementation in the industrial domain, manifest themselves through algorithmic, technological and methodological challenges, or through a synergistic combination of these aspects.

On the technological front, research continues to push the boundaries of foundational Industry 4.0, technologies, such as the IoT, AI, ML, advanced robotics, AR and VR, and blockchain. Researchers explore ways to improve connectivity, energy efficiency, data security and computational capacity, while making these technologies more accessible to companies of all sizes. With the increase in connectivity, cybersecurity becomes a crucial research domain. Researchers work on advanced methods of risk analysis, securing industrial networks and protecting sensitive data against cyberattacks, developing dynamic solutions adapted to evolving threats.

Cyber-physical systems (CPS) are at the heart of Industry 4.0,, integrating physical processes with their digital twins. Research focuses on improving the bidirectional interaction between physical and digital components. It is equally important to establish standards that facilitate the integration and interoperability of CPS into value chains without additional development. With the proliferation of data generated by digitalized industrial systems, research aims to develop computational models including the use of AI and ML to improve decision-making and optimize production processes.

On the methodological front, researchers propose and experiment with new frameworks to guide companies wishing to engage in a transformation process. The proposed methods, organized into several phases, constitute a structured roadmap. Furthermore, they often provide recommendations in terms of prerequisite requirements and offer an overview of the necessary environment to achieve targeted functionalities.

These frameworks are designed to be sufficiently flexible and adaptable to various industrial sectors while being structured enough to provide clear guidance. Generally, these frameworks begin with an assessment of the company's current state, identifying technological gaps, available skills and specific digitalization needs. Furthermore, they offer specific recommendations regarding prerequisite requirements, particularly in terms of control architecture evolution or IT infrastructure. These frameworks also provide an overview of the technological and human environments necessary for the targeted functionalities to become operational. For example, they may emphasize the importance of creating multidisciplinary teams to facilitate technological integration into existing production processes, or the imperative of developing a data strategy to optimize the collection, analysis and use of information generated by connected systems.

This research work, the purpose of which is to produce industrial methods, helps reassure companies wishing to undertake digital transformation progressively. Indeed, Industry 4.0 is often perceived as a disruptive revolution and can arouse legitimate apprehensions regarding its adoption. Thus, these approaches encourage a holistic approach, emphasizing the alignment of targeted objectives with the development trajectory and market needs, thereby ensuring that the transition to digitalized industrial systems translates into tangible productivity improvements.

Furthermore, researchers also examine how Industry 4.0 transforms economic and work models, studying the impacts on employment, skills development, work organization and relationships between companies and their stakeholders. They explore approaches for a successful transition toward agile collaborative models. Moreover, there is a growing awareness of the social and ethical implications of Industry 4.0. Research in this domain also aims to understand and address questions related to equity and technological inequalities, proposing an approach for responsible and inclusive development of technologies. Similarly, sustainability and the circular economy represent an expanding research domain. This includes resource optimization, waste and emissions reduction, as well as the promotion of more sustainable production practices.

On the algorithmic front, the objective is to develop dedicated algorithms, including aspects of analysis, decision-making, autonomy, prediction, as well as communication with different platforms. Their roles consist of complementing existing solutions for specific needs that were not initially foreseen or that can further optimize industrial processes.

These algorithms integrate different computational strategies based on AI, such as ML, deep learning and optimization. These approaches make it possible not only to analyze and process vast quantities of data in real time but also to make decisions autonomously, adapted to the specific situations encountered. For example, in production performance monitoring, algorithms can identify a combination of behaviors indicating early-stage degradation. This is made possible by analyzing data from IoT sensors and using complex models to anticipate performance changes.

Furthermore, these advanced algorithms must interact with different platforms and devices within the industrial ecosystem. Through the use of standardized communication protocols and interoperable data formats, they enable fluid synchronization and coordination between machines, applications and production control systems, thereby promoting increased visibility across the entire production process.

Research in this domain is an undertaking that requires multidisciplinary expertise, encompassing the fields of AI, software engineering, networks, industrial processes and embedded systems, in order to provide intelligent, scalable and interoperable solutions that meet the challenges of digital transformation in industry.

In summary, Industry 4.0 continues to grow thanks to scientific research and corporate development initiatives, with a constant flow of new discoveries and innovations.

Initially centered on the manufacturing sector, the benefits of Industry 4.0 now extend to various other domains such as agriculture, health, logistics, etc. This is confirmed by a scientific output that progresses year after year (see Figure 1.1). Indeed, an analysis conducted via Scopus over the last 10 years using the four keywords, *Industry 4.0*, *smart manufacturing*, *cyber-physical systems* and *IoT in manufacturing*, shows an upward trend that is tending toward a stabilization in the number of articles per year. Furthermore, the United States, China and Germany stand out as the main contributors to this scientific output.

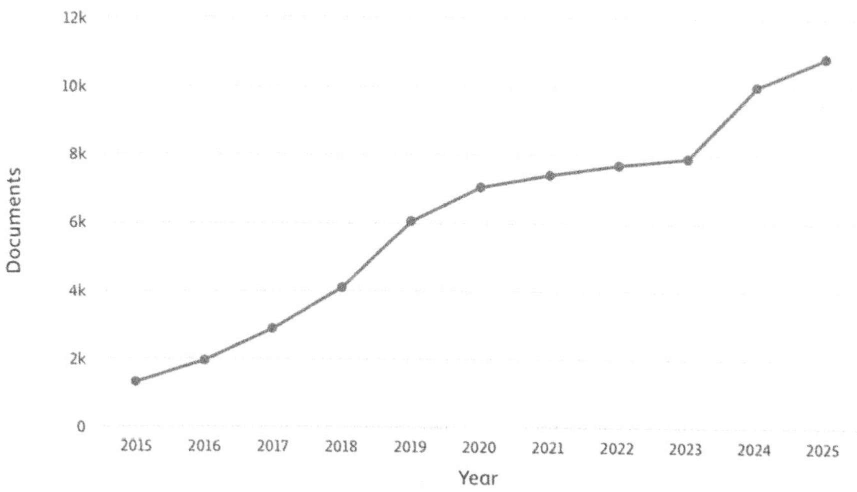

Figure 1.1. *Analyze Scopus*

Companies and governments are increasing their investments in research and development (R&D) related to Industry 4.0. These investments primarily target the integration of AI into industrial systems, the Industrial Internet of Things (IIoT), advanced manufacturing execution systems (MES), cloud manufacturing, cobotics and cybersecurity, with growing attention to societal and environmental aspects. The global Industry 4.0 market is booming. With forecasts indicating that it could reach several hundred billion dollars by 2030, the state of research is expected to evolve significantly in the coming years.

1.4. A global strategy

The adoption of Industry 4.0 must align with the company's overall vision and include clear objectives. These objectives can also be expressed in terms of desired functionalities, such as flexibility, predictive capability, portability, or increasing the autonomy of the production system or adapting to its new digitalized ecosystem. The transition toward digitalization of industrial systems requires a well-thought-out overall strategy to maximize its benefits and minimize associated risks. This strategy should cover not only the technological aspect but also cultural transformation, organizational change and employee training. To this end, the literature offers the main steps that could structure a transition approach.

The first step consists of assessing current digital maturity. This involves analyzing processes, existing technologies and company culture. Understanding where the company stands in terms of digitalization and automation makes it possible to establish the starting point. Next, define precise objectives, which can be expressed in terms of expected functionalities. They must be specific, measurable, achievable and realistic. A detailed assessment of current processes, systems and skills allows for evaluating the effort and time necessary to achieve the objectives or, conversely, to redefine the priority levels of each objective. The final element of this step concerns planning that takes into account contingencies and uncertainties that may hinder this industrial transition, particularly in terms of rapid technological evolution, dependence on solution suppliers, concerns expressed by employees, protection of data and know-how, etc.

The second step aims to select and prepare for technological integration by relying on methods developed for this purpose. The integration of Industry 4.0 technologies can conceal intrinsic difficulties due to their interconnection and lack of interoperability. Indeed, the majority of industrial equipment is not natively interoperable, hence the need to develop specific "gateway"-type applications that promote reliable and sustainable communication of data and actions. Technological components are identified according to their direct or indirect contribution to the objectives. Here are some examples:

– the Internet of Things (IoT): used to connect machines and systems;

– analytics: to analyze data and optimize processes;

– artificial intelligence (AI): to improve decision-making;

– cybersecurity: to protect critical data and infrastructure;

– 3D printing: for flexible local production;

– augmented reality (AR): for assistance in maintenance and training processes.

The third step concerns training and skills development. Training personnel to acquire new skills related to the digitalization of industrial systems constitutes an essential success factor. This facet of the strategy is far from negligible. The reason is that this transformation is accompanied by a complete overhaul of the tasks assigned to employees. For example, maintenance personnel should be capable of identifying, collecting and analyzing data via AI platforms in order to propose a dynamic maintenance plan aligned with the desired availability objectives. The same applies to the production manager, who will have to interact with advanced, autonomous and intelligent control systems. This underscores the importance of promoting a culture of experimentation, where failure is seen as a learning step.

The fourth step consists of defining pilot projects to test concepts and technologies, enabling companies to evaluate them in a controlled environment before proceeding to large-scale deployment. These projects constitute an effective means of demonstrating the added value of new technologies and methodologies. By presenting tangible results, it will be easier to convince internal and external stakeholders of the value of the digital transition. Pilot projects also make it possible to initiate a collaborative dynamic, involving employees, partners and customers in the experimentation and evaluation process. This participatory approach contributes to better acceptance and adoption of changes. By working directly with new technologies, personnel gain practical experience will be invaluable for the long-term success of the digitalization initiative.

The fifth step of the transformation toward more digitalized and innovative production systems consists of joining or creating innovation ecosystems that aim to promote a dynamic sharing of knowledge, resources and technologies. They involve a strategy of openness and cross-functional collaboration with external entities such as startups, universities and other research institutions. This approach is essential for sustaining the transition over time. Indeed, increasing technological complexity requires specialized skills that are often rare or under development. The lack of experts in cutting-edge fields can slow down transformation initiatives. Furthermore, the technological landscape is evolving at an unprecedented pace, with continuous advances that can hardly be monitored in isolation. By joining or creating innovation ecosystems, companies can benefit from a greater diversity of expertise and perspectives. Moreover, close collaboration with external partners, such as technology startups and academic institutions, can provide privileged access to emerging talent and customization services for company-specific technological solutions.

The example of Universal Automation is revealing. This independent, nonprofit association oversees the implementation of an open-source execution engine dedicated to industrial automation, based on the IEC 61499 standard. At its creation in November 2021, the association had nine founding members from user and supplier communities. Today, it brings together more than 90 members, including automation users, technology suppliers, universities, institutions and startups.

The sixth step of the digital transformation process is fundamental to ensuring the success and sustainability of the initiative. It consists of establishing and monitoring key performance indicators (KPIs) that serve as benchmarks for measuring progress and the effectiveness of new technologies and processes put in place. This enables companies to quantify the benefits of their digital transformation and justify the investments made. Relevant KPIs vary according to each company's specific objectives and can cover a wide range of areas, including, but not limited to, operational autonomy, production flexibility, customer satisfaction and availability rate.

The next phase is the adoption of a continuous improvement approach to optimize the performance of processes and technologies. This may involve the use of advanced computational resources such as data analytics, AI and ML to analyze performance, identify trends and predict future events. In summary, the combination of KPI measurement and continuous improvement represents a powerful strategy for maximizing the benefits of the digital transition and for ensuring a constant level of effectiveness within the organization.

In summary, an overall strategy for Industry 4.0 must be shared and progressive. The key to success lies in a balanced approach that combines technology, people and processes, while maintaining vigilance over security and long-term sustainability.

1.5. Business models

Industry 4.0, with its integration of advanced technologies, is shaping new business models. These models capitalize on digitalization, connectivity and data analytics to transform production, distribution and consumption methods. This leads to new business models where value is created not only by the product itself but also by the ability to meet specific customer needs. Below is an overview of some emerging business models.

The "As-a-Service" model represents a significant transformation in the way companies access and use technologies and equipment. Instead of making a substantial upfront investment to acquire assets, companies can now access what they need via a subscription, thereby transforming capital expenditures (CapEx) into operational expenditures (OpEx). This approach offers considerable financial and operational flexibility, adapting expenses to real needs and company growth. This model can extend to heavy equipment as equipment-as-a-service (EaaS), software as software-as-a-service (SaaS), and even platforms as platform-as-a-service (PaaS), thus reducing initial costs and linking payments to usage or expected performance.

Within the framework of EaaS, companies no longer purchase heavy or specialized equipment, but pay to use it on a subscription or usage basis. This can concern industrial machinery, vehicles or even high-tech tools. This approach enables suppliers to monitor the behavior of their products during the usage phase through the exploitation of data collected and analyzed remotely. In this configuration, the supplier markets access to the product to companies while assuming maintenance operations and end-of-life management. This approach encourages industrial equipment suppliers to pay particular attention to the durability of their products. Through the analysis of lifecycle data, it becomes possible to extend the valorization of equipment over a longer period and in a flexible manner, thereby making subscription costs more competitive.

Furthermore, the combination of 3D printing, AI and automation makes it possible to produce customized items, reducing dependence on global supply chains. This fosters the emergence of more resilient business models, while reducing logistics costs and environmental footprint.

The emergence of the "platform economy" model is accelerating with the advancement of digitalization of industrial processes and advanced control systems, which now make it possible to measure, over a time sequence, the load rate of industrial equipment. In the manufacturing sector, this translates into the creation of a new market for unused production capacities, offering companies the opportunity to monetize excess capacity to other companies seeking to temporarily increase their production. Digital platforms play a key role in this model by facilitating dynamic and efficient exchanges between manufacturers, thereby optimizing resource utilization and opening new revenue streams.

The analysis of massive data, particularly data arising from product usage, enables companies to better understand consumer behaviors and preferences. This allows them to adjust their business models according to market evolutions, thanks to intelligent control systems that offer personalized products, optimize resource utilization, reduce waste and promote the use of renewable energies.

The integration of advanced technologies helps optimize inventory and improve demand planning. Furthermore, the use of blockchain offers unparalleled product traceability throughout the supply chain. This can be used to guarantee product authenticity, reduce intermediaries, paving the way for new business models based on trust, transparency and direct-to-consumer distribution.

In conclusion, companies that adopt and adapt to more sustainable and environmentally respectful consumption trends are those that will remain competitive and prosperous in the new global economic landscape. These business models are made possible by technological advances and offer opportunities for growth, innovation and competitiveness in a constantly evolving market.

1.6. Data transformation

In Industry 4.0, data transformation plays a central role, acting as the fuel that powers all intelligent and automated processes. Their transformation begins first with a data collection process. Sensors integrated into machines, equipment and even finished products collect real-time data on their status, performance and environment. Similarly, cyber-physical systems generate and collect data from physical processes. Connectivity between machines, systems and actors (suppliers, customers) enables large-scale data collection, offering multi-scale views: first local for operations control, then global for the entire value chain. Finally, business information, such as production or maintenance plans, also represents a data source to consider.

Collected data are analyzed via advanced analytics techniques to extract insights, trends and predictions. AI and ML transform data into concrete actions, enabling systems to learn from their environments and optimize their operations without direct human intervention. Data transformation also includes their conversion into visually comprehensible formats, allowing decision-makers to quickly grasp complex situations.

In terms of applications, they can be grouped into two categories: real time and deferred time. For deferred-time applications, data interpretation makes it possible to predict failures before they occur. Consequently, maintenance planning becomes dynamic, thereby reducing downtime. Furthermore, thanks to in-depth processing of accumulated data, it becomes possible to identify cause-and-effect relationships that would not be evident without this analysis, such as a gradual decline in quality or deterioration of precision, phenomena for which direct human observations provide no explanation. This type of a posteriori analysis contributes to improving the overall performance of digitalized production systems.

In real time, collected data make it possible to increase the autonomy of production flow control systems via dynamic scheduling of operations. This increases, on the one hand, the capacity to customize mass production, and, on the other hand, the ability to propose and execute a new scheduling of operations in the event of contingencies, offering continuity of flow and products adapted to specific consumer needs.

With the increase in volume and variability of collected and shared data, the reliability of analyses takes on capital importance. The ability to make informed decisions and optimize production processes invariably rests on the quality and integrity of collected data. This imperative raises the need to implement rigorous data verification and cleansing mechanisms, to ensure that the information used in the decision-making process is both accurate and reliable.

Data verification involves detecting errors, anomalies or inconsistencies that may be concealed within collected data. This is crucial because in a highly integrated industrial production environment, a minor error in data can trigger a chain of costly failures or erroneous operational decisions. Data cleansing, which follows the verification phase, consists of rectifying or eliminating data presenting dubious singularities to ensure that the resulting analysis is based on reliable information. These steps are all the more crucial in Industry 4.0 where decisions must often be made quickly, based on massive volumes of data generated in real time.

However, transforming collected data into actionable insights is a process that requires specialized skills, far beyond the mere ability to manipulate data analysis tools. First, in-depth expertise in data analysis is essential. Professionals must not only understand the underlying statistical principles but also know how to apply these principles to heterogeneous datasets in order to extract useful information that will support strategic decision-making.

Second, AI and ML make it possible to analyze massive quantities of data in real time, identify complex trends and patterns, and suggest predictions or recommendations with precision. Mastery of these technologies is therefore decisive for fully leveraging the potential of data in the continuous improvement of production processes.

Finally, effective management of information systems is a key skill in a digitalized industrial dynamic. It involves implementing an adapted technological infrastructure for the secure collection, storage and analysis of data. In the case of massive data, the concept of a data lake can be a reliable solution for preserving raw data in their native format until they become useful. Unlike traditional storage systems, which store data in structured files or folders, data lakes are designed to store large volumes of data in various formats, such as texts, videos, images, sensor data, etc., without requiring prior organization or structuring. This approach offers remarkable flexibility by enabling companies to save all their data in a single place, thus facilitating access and analysis.

One of the main advantages of a data lake is its capacity to support big data analytics and machine learning at large scale. Companies can explore vast data reservoirs to discover valuable insights, detect trends or even predict future behaviors without the constraints imposed by the need to structure data before storage. However, effective exploitation of a data lake requires governance mechanisms to prevent it from transforming into a data swamp, i.e. a data lake where the absence of management and classification renders data difficult to find and exploit, ultimately of low value.

Pertinent management of information systems guarantees not only data security but also their integrity and availability for real-time or deferred-time analyses. Within the framework of Industry 4.0, transforming data into reliable insights requires a combination of specialized skills in data analytics, AI and information systems management. Data quality and integrity, as well as the ability to transform data into actionable information, are essential for supporting operational decisions and optimizing production processes in an increasingly autonomous industrial environment.

2

New Functionalities

The development of Industry 4.0, and by extension Industry 5.0, encourages companies to invest further in the digitalization of their manufacturing processes. Similarly, educational institutions are launching projects to acquire teaching materials adapted to the future skills sought by industrial partners. This raises the question of transition approaches toward digitalized industrial systems or, even more, the design of new production lines natively integrating the digital dimension. To achieve this, one of the first steps consists of translating expected or desired performances into functionalities, whether they pertain to Industry 4.0 or Industry 5.0.

The precise definition of the sought-after functionalities, from the very beginning of the transition or design process, constitutes an essential prerequisite to guarantee that the final system effectively responds to the company's evolution needs. This step makes it possible not only to target performance objectives but also to define the technological bases and required capacities for smoother integration and better acceptance by end users. Thus, this chapter presents a synthesis of the functionalities that can characterize digitalized production systems. The upstream selection of functionalities makes it possible to better channel investments.

The new functionalities offered can be organized into three main categories. The "preparation" and "control" categories are considered successive phases in a manufacturing production logic. The "cross-functional" category covers the entire manufacturing process phase and can extend to upstream and downstream phases. It constitutes an essential link in the digital chain that traverses the different stages of the lifecycle and explores the data generated. Data processing and analysis call upon computational resources, whether local or remote, with a view to producing information that can particularly optimize design, industrialization or industrial production.

Before presenting the different functionalities, we first introduce the three categories mentioned above. As indicated in Figure 2.1, the functionalities are grouped into three categories. Similarly, each functionality can group several subfunctionalities whose interpretation may evolve or, in certain cases, change categories if we consider another point of view. A synthesis of the main functionalities drawn from various bibliographic or documentary sources is presented in Figure 2.1.

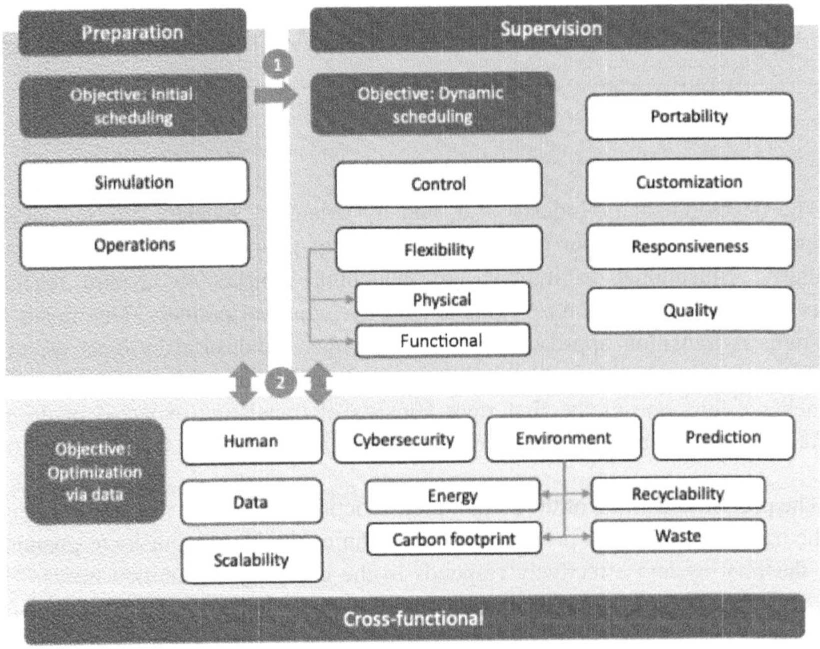

Figure 2.1. *Summary of functionalities 2.1*

2.1. "Preparation" category

The first category, known as "preparation", aims to prepare production operations as well as their initial scheduling for deployment in the production phase. This phase, also called the "industrialization phase", has fundamentally evolved in the era of digitalized production systems. Indeed, 4.0 operations are complex because they encapsulate both a layer of conventional information, useful for executing machine sequences, and a layer of information necessary for bidirectional dialogue with their digital environment. Generally, this phase requires the involvement of multidisciplinary experts, particularly if the goal is to develop parametric operations that will be added to the library of standardized operations.

This category brings together two main functionalities: "operations" and "simulation".

2.1.1. *Operations*

The functionality described here plays an essential role in the supervision of modern production systems, particularly with regard to the manufacturing execution system (MES). It aims to standardize and parameterize elementary production entities, the basic building blocks of manufacturing processes, by assigning them various adjustable parameters. This approach offers greater autonomy and flexibility to the MES, which is crucial in a dynamic production environment.

Elementary production entities can be understood as the fundamental building blocks in a manufacturing process. They are designed to be both standardized and parametric. Standardization ensures that components are interchangeable and compatible on a large scale, while parameterization makes it possible to adjust production variables, such as the materials used, environmental conditions or final product specifications.

By integrating these parametric entities, the MES acquires the ability to adjust parameters in real time, in response to changing conditions or unforeseen events on the production line. For example, if a machine detects an anomaly in the raw material, the MES can instantly modify the production parameters of the affected elementary entities to compensate for this variation and maintain the production flow. This flexibility makes the production process more resilient and efficient, minimizing downtime.

The design of elementary production entities represents a complex challenge that mobilizes skills from several disciplines. This process is essentially divided into two main layers. First, the parametric programming layer, known as G-code, encompasses the code used to direct machine tools or robots. This G-code is designed to be adjustable, thus offering flexible programming adapted to the specific needs of production. Second, the bidirectional communication layer is essential for integrating the entity into a broader digital environment. This layer facilitates exchanges between the production entity and the rest of the manufacturing system, including other machines, databases, as well as monitoring and supervision software. This bidirectional communication is crucial for real-time feedback and the continuous adjustment of operations.

In summary, the creation of standardized and parametric elementary production entities enriches the MES with an adaptive capacity that is vital for responding effectively to the demands of flexible production. This functionality is not limited to increasing flexibility alone but also contributes to the efficiency of industrial processes. It requires a deep integration of technical, computing and production management skills, underscoring the importance of a holistic and integrated approach in the design of modern manufacturing systems.

2.1.2. *Simulation*

The immersive simulation functionality in the industrial domain aims to virtually reproduce the behavior of production operations as well as the flow of production sequences. The main objective is to enable engineers and supervisors to visualize and interact with production processes in a virtual environment before their physical implementation. This approach makes it possible to test and validate production processes and flows collaboratively, thus ensuring that everything functions according to the planned specifications before the actual launch.

The advantages of immersive simulation are numerous. It offers the possibility of validating manufacturing processes upstream, allowing teams to identify potential problems and ensure that everything conforms to expectations. This includes verifying compliance with quality, safety and performance standards. Similarly, it enables the visualization of scenarios calculated by the MES. Decision-makers can thus visualize how the MES plans to orchestrate operations and, if necessary, adjust sequences to optimize workflows.

Furthermore, it contributes to the detection of malfunctions. Indeed, by simulating production operations, it is possible to detect failures at the level of elementary production entities before they cause problems in a real production environment. This makes it possible to identify flaws in the configuration of machines or in the interaction between different cyber-physical systems (CPS) and the MES. Finally, immersive simulation can also highlight dialogue problems between cyber-physical systems and the MES. Identifying these intersystem communication issues at an early stage allows teams to work toward their resolution before production is affected, thus ensuring smooth and efficient coordination between the various components of the production system.

To implement this simulation, production data, machine parameters and MES scheduling algorithms are integrated into a virtual model. Engineers then use virtual or augmented reality tools to interact with the simulation, which enables adjustments to be made and different production scenarios to be tested in a safe and controlled manner.

In summary, immersive simulation in digitalized production processes offers significant advantages by enabling detailed visualization and collaborative, potentially remote, validation of production processes. This reduces the risk of errors and limits malfunctions.

2.2. “Supervision” category

The complexity of digitalized production systems, particularly those designed on the basis of a distributed architecture, has given rise to a new supervision logic based on an interweaving of intelligent, autonomous systems capable of processing a mass of real-time data. This approach marks a break from previous systems and defines new relationships with production supervisors and advanced supervision systems. The role of supervisors or managers is evolving radically. They will be required to intervene in upstream phases to define precedence constraints and optimization priorities associated with each series of components, such as minimizing energy consumption, prioritizing execution time or producing at lower cost. Similarly, they are present during production execution to assess performance evolution or in the event of unforeseen adverse events requiring one or more human interventions. More generally, they remain the guarantors of the overall coherence of the production system.

Furthermore, the digitalization of production systems has transformed the traditional approach to operations scheduling, moving from a static, predefined model to a dynamic and adaptive system. This evolution is particularly evident in the way production operations are planned and executed, adapting to changes in the manufacturing environment. Dynamic scheduling algorithms make it possible to manage unforeseen events in real time, activate diagnostics in the event of malfunction, inspect the production line, secure dynamic components in case of intrusion and make necessary adjustments when performance deviates.

Algorithms dedicated to dynamic scheduling use continuous data from IoT (Internet of Things) sensors and other integrated sources to readjust production plans or reparameterize operations based on real-time changes in the production chain. This includes demand variations, execution delays in certain phases, equipment breakdowns or changes in operator availability. This ability to adjust plans quickly minimizes downtime and optimizes resource utilization. Furthermore, they also enable greater product personalization without compromising production efficiency. Systems can be programmed to adjust production lines to manufacture different product types according to specific customer needs, while maintaining an optimal production flow.

Technologies such as artificial intelligence (AI) play a crucial role in dynamic scheduling. They make it possible to analyze upstream data to optimize the sequencing of operations in real time, identify anomalies and suggest adjustments. They also offer supervisors the possibility of prioritizing scheduling dynamics based on criteria such as cost, energy or time.

The interface between human operators and automated production systems is enhanced through interactive dashboards and intuitive user interfaces. Operators can receive real-time alerts, visualize scheduling changes and intervene manually if necessary. This human–machine collaboration increases responsiveness in decision-making and strengthens operational efficiency.

Finally, the integration of dynamic scheduling with supply chain management systems ensures that production flows are well coordinated with logistics, inventories and distribution. This makes it possible to reduce excess inventory, improve delivery times and respond more effectively to market fluctuations.

In summary, dynamic scheduling in digitalized production systems is a sophisticated functionality that reflects the complexity and dynamism of modern production environments. It illustrates how advanced automation and embedded intelligence can radically transform traditional manufacturing practices, contributing to the successful execution of production plans in real time. The "supervision" category brings together functionalities that collectively contribute to the dynamic management of operations, the availability of the production line and product conformity.

2.2.1. *Flexibility*

The digitalization of manufacturing processes has brought considerable advances in the way factories operate, with a particular focus on increasing flexibility. The latter has become an essential characteristic, enabling production tools to adapt quickly to evolving market demands. To better understand this flexibility, it is useful to examine it from two main perspectives: physical and functional.

Physical flexibility concerns the ability of a factory to physically reconfigure its production resources, notably machines and cyber-physical systems (CPS), in response to production plan requirements. This reconfiguration can be activated by the MES or by supervisor decision. It depends largely on the modular design of the workshop, which allows rapid modifications to the layout of production equipment. For example, mobile components such as autonomous robots can be moved and reprogrammed to perform different tasks, ranging from assembly to quality inspection, depending on current production needs.

On the other hand, functional flexibility refers to the ability of a production line to manufacture various components within the same production flow without requiring prolonged stoppages to reconfigure equipment. This capacity is crucial for factories that wish to maintain continuous production while responding to a diverse range of products or components. In such an approach, the different components follow a predetermined route through a series of CPS, where they undergo various operations such as machining, inspection or marking. The key to this flexibility lies in a distributed and interconnected configuration of the workshop and in the level of robotization.

To implement these types of flexibility effectively, the factory architecture must be designed with high modularity and interconnectivity. Production management systems, such as the MES, must be integrated with advanced communication and control capabilities to coordinate production smoothly and responsively. Furthermore, staff training is decisive, ensuring that operators can effectively reconfigure and optimize production resources according to changes in demand.

In summary, organic (physical) and functional flexibility in digitalized manufacturing processes enable companies to leverage digitalization. By rapidly adapting production to fluctuating operational conditions, factories can maximize the workshop utilization rate, an indicator that measures the effective use of production capacities relative to their maximum capacity. These aspects of flexibility are among the key pillars of Industry 4.0.

2.2.2. *Control*

The control functionality in digitalized production systems is necessary to ensure agile and reactive management of manufacturing operations, as it is no longer possible to entrust this mission to human operators due to heterogeneous and rapid production involving a significant quantity of data, parameters and variables. Thus, this control is primarily orchestrated by the manufacturing execution system (MES), a crucial tool that enables factories to adapt dynamically to unforeseen events or changes in priorities. The ability of the MES to reorganize and reparameterize operations in real time is a major asset in the continuous optimization of production. This functionality is characterized by several capabilities

The autonomy of the MES is manifested in its ability to reorganize the scheduling of operations based on real-time data and feedback from ongoing processes. This functionality makes it possible to react quickly to interruptions or variations in workloads, thus minimizing downtime and maximizing the utilization rate of industrial equipment.

Similarly, the MES optimizes production according to priority criteria such as cost, lead time and energy consumption. It adjusts production parameters to ensure the best balance between these factors, thus ensuring optimal exploitation of available resources. Furthermore, the MES has the ability to make operational decisions in the event of specific occurrences. It can, for example, decide to reroute certain operations to other machines in the event of breakdown or stoppage for maintenance.

The MES is also capable of conducting diagnostic operations to identify and analyze the causes of defects or delays. This functionality is essential for maintaining a high level of quality and for preventing problem recurrence.

Finally, the smart dimension of the MES lies in its ability to generate solutions adapted to complex problems. This includes dynamic adaptation of workflows, optimization of machine parameters and minimization of the production system's energy consumption.

For the MES to effectively cover all these attributes, it is imperative that it has access not only to historical data but also to data generated in real time by the various points in the factory. Interoperability between the MES and other systems such as enterprise resource planning (ERP), CPS and IoT equipment is crucial. This integration ensures an updated view of production status, enabling the MES to make informed decisions.

Furthermore, the ability of the MES to consider external factors, such as variability in market demand, supply chain conditions or even weather conditions, is essential for maximum responsiveness. This allows the system to adapt proactively, not only to internal conditions but also to those coming from the external environment.

In summary, the MES plays a pivotal role in the digitalization of manufacturing processes, acting as the brain that supervises, optimizes and adapts production to respond effectively to evolving needs and unforeseen situations. Its implementation and continuous development are therefore essential to the operation of integrated and digitalized workshops.

2.2.3. *Personalization*

Mass customization consists of personalizing products in an individualized manner and on a large scale. This approach, made possible by significant technological advances, makes it possible to meet specific consumer requirements while maintaining production efficiency.

To facilitate this mass personalization, it is essential that all data generated by machines, the production process, the quality system and IoT sensors are not only collected and archived but also tagged for each component. This accumulation and analysis of data enable detailed tracking of each part through the production system, offering complete traceability and advanced quality control capabilities.

The MES, whose traditional role is to orchestrate production operations, must now manage increased levels of complexity by processing information on a per-part basis. To this end, the MES must adapt production orders in real time to incorporate personalized specifications while optimizing resource allocation. This functionality relies on the intrinsic flexibility of the production system, including the ability of machines to change configuration quickly, the use of reprogrammable collaborative robots (cobots) for different tasks and the implementation of agile logistics systems capable of managing a variety of incoming materials and components.

In conclusion, mass personalization is an advanced functionality and draws upon other functionalities. It requires facilitated access to data and advanced coordination between the MES and other software/applications within the production system. The effectiveness of this personalization relies on the ability to integrate these technologies into a seamless process while maintaining the conformity of manufactured products.

2.2.4. *Responsiveness*

Responsiveness in the context of industrial production can be analyzed through two main levels: the "workshop" level and the "supervision" level. These two distinct levels interact to optimize the system's response to variations and unforeseen events, relying on a reconfigurable technological infrastructure and intelligent algorithms.

At the workshop level, responsiveness is directly influenced by the infrastructure's ability to adapt and reconfigure the geographical layout of certain equipment. This requires the integration of complex technologies such as autonomous mobile robots equipped with manipulator arms or robots capable of moving certain industrial equipment autonomously. These robots are also equipped with sophisticated sensor networks, including cameras, which enable the acquisition of environmental and operational data. This data collection allows rapid adaptation to physical variations in the workshop. Furthermore, the workshop's ability to produce locally, via 3D printing, certain components necessary for product personalization or the rapid replacement of faulty parts, is another aspect that characterizes responsiveness.

The supervision level concerns how data collected at the workshop scale is used to make the production system more responsive. The MES relies on analysis algorithms to detect any deviation, loss of performance or early signs of material degradation. Thanks to embedded analytical capabilities, the MES can make proactive decisions, such as the visual inspection of a specific area, by requesting mobile robots to perform a detailed local check. It can also dynamically control resources and operations, adapting the production flow according to unforeseen events. For example, in the event of an unexpected machine stoppage, it immediately redirects the flow to available machines.

The full web interface of the MES is another element promoting responsiveness. This interface makes collected and analyzed information accessible in real time on various devices, such as smartphones and tablets, enabling all stakeholders to intervene quickly and with full knowledge of the facts, regardless of their location. The ability to access and react to information in real time through connected devices significantly increases the company's responsiveness.

Thus, responsiveness in a modern industrial context is not limited to the speed of response but encompasses an integrated and intelligent management of production that combines advanced technology, data analysis and fluid communication in order to maintain overall performance.

2.2.5. *Portability*

The MES offers various degrees of control and interaction according to specific needs. It typically proposes two main configurations: a "limited" option and an "extended" option, each designed to adapt to different levels of operational requirements.

The "limited" option of the MES is primarily focused on providing remote access to essential production information. It allows authorized users, whether internal to the company or external, to consult production monitoring indicators and real-time alerts. This is achieved through connected user interfaces that facilitate the visualization and tracking of essential steps. This connectivity improves transparency and can help strengthen collaborations and relationships with stakeholders by offering a direct and current overview of ongoing operations. Furthermore, this option may include the possibility of sharing certain critical information via a secure interconnection between the MES of different entities, thus promoting an integrated and responsive supply chain.

The "extended" option of the MES goes beyond simple process visualization to encompass remote operational control capabilities. This option allows users to program and manipulate specific aspects of production directly through the MES. For example, it can authorize the remote command of a mobile robot equipped with a camera to inspect anomalies on the production line. This may include video supervision of a situation and, if necessary, the initiation of corrective actions such as machine parameter adjustments or modifications to the production process, all carried out remotely by qualified and authorized personnel.

Furthermore, this option also makes it possible to act on the prioritization parameters of the MES, thus adjusting the scheduling of production sequences according to evolving needs or specific situations encountered.

Nevertheless, the implementation of the extended option is complex and requires particular attention to security and reliability. Securing the production line and protecting personnel on site are imperative, especially when remote interventions can affect ongoing machines and processes. The management of potential conflicts between different instructions sent to the MES is also a significant challenge. Systems must be designed to prioritize commands effectively and manage contradictions in a way that avoids errors and accidents.

The distinction between the "limited" and "extended" options of the MES illustrates the flexibility and adaptability of these systems in the contemporary industrial context. Each option offers varied levels of interaction and control, enabling companies to choose the solution that best corresponds to their specific needs in terms of production monitoring, supervision and responsiveness.

2.2.6. *Quality*

In the industrial world, the objective of quality has not evolved. The quality function aims to verify the conformity of products, processes and equipment using predefined instruments and protocols. The digitalization of industrial processes has profoundly transformed the quality approach and has given rise to the "Quality 4.0" functionality, which is more integrated into the digitalization of industrial processes.

"Quality 4.0" is driven by two aspects. The technological aspect, through the development of contactless inspection means and their software for acquiring and processing collected point clouds. The combination of this technology, increasingly miniaturized and powerful, with the ability of robots to carry this equipment close to components during manufacturing has led to undeniable advances. Indeed, inspecting components within machines makes it possible, in the event of deviations, to correct the process or interrupt it. This is only possible if the production system benefits from operational flexibility that allows the MES to instantly modify sequences and/or the scheduling of production operations.

The second aspect concerns data at the scale of the production system. Indeed, the implementation of AI can reveal correlations between observed deviations and the behavior of the production process. By way of example, AI can draw our attention to the presence of nonlinear behavior in deviations attributed to a given machine, observed at the moment when another nearby machine is solicited, thus generating vibratory propagation that disrupts surrounding machines. Thus, this information enables the MES to consider this constraint in dynamic operations scheduling, as a compromise between performance stabilization and deadline compliance.

The ability of the MES to monitor product conformity with respect to required quality is made possible by the digitalization of the production system, contactless inspection means and the performance of embedded intelligent algorithms.

2.3. "Cross-functional" category

The third category, termed "cross-functional", generally relies on data from the production system and aims to optimize manufacturing processes, or even the redesign of the component. In real time, the computational dynamics of data will optimize or influence the unfolding of the manufacturing process. In deferred time, data analysis can produce information useful for the repair phase, particularly in terms of phenomenological interaction between the production line and the components that move through it. This category brings together several functionalities: prediction, cybersecurity, environmental impact and the human dimension in the manufacturing process.

2.3.1. *Prediction*

The "prediction" functionality makes it possible to predict undesirable events or deviations based on advanced analysis of collected data. Undesirable events are generally stoppages or loss of performance due to breakdowns or degradation. The objective is to ensure operational continuity of the production chain, optimize maintenance and forecast the availability of spare parts, particularly for companies operating in just-in-time flow.

To achieve this objective, there are two approaches. The first consists of using AI resources to construct a model of normal machine operation without using historical data. Once constructed, it is considered that any deviation from the normal model may constitute an anomaly, which is then reported via the MES for assessment. This approach makes it possible to detect early warning signals of failures or degradation. This model activates in real time and requires a high-performance IoT device, particularly for vibratory or acoustic signals.

The second approach is based on the analysis of archived data in order to detect correlations between the occurrence of breakdowns or deviations and the behavior of the production process. The objective is to identify triggering factors and transform them into operational constraints that the MES should consider, both in the preparation phase of production plans and also in the operations execution phase. For example, a temporal correlation may be detected between deviations observed in the quality of several components and the startup of a specific machine that generates significant vibrations. Thus, this analysis results in a new constraint that the MES will take into account in operations scheduling for components that require manufacturing precision.

2.3.2. *Cybersecurity*

Cybersecurity, in the context of Industry 4.0, is no longer an option but a strategic necessity essential to integrate from the initial phases of any industrial digitalization project. Indeed, as industry moves toward interconnected production systems, vulnerability to cyberattacks increases, making digital security issues central to the design and operation of modern industrial systems.

The transition to digitalization brings with it a considerable increase in access points and interfaces susceptible to being exploited by cybercriminals. This hyperconnectivity certainly facilitates communication and automation of industrial processes, but it also opens potential breaches for malicious intrusions. Consequently, the integration of cybersecurity from the system design stage, known as Security by Design, becomes crucial. This principle ensures that security aspects are taken into account from the beginning of system development, rather than being added as an additional layer after the fact.

Cybersecurity primarily aims to protect digital resources that include sensitive data such as industrial secrets, production plans, employee personal data and customer-related information. It seeks to guarantee three fundamental principles: confidentiality, integrity and availability of data. Confidentiality ensures that access to data is limited to authorized individuals. Integrity prevents unauthorized modifications of data. Availability guarantees that information and systems are accessible to legitimate users when necessary.

To achieve these objectives, companies must adopt a combination of technological solutions and best practices. This includes the use of advanced firewalls, intrusion detection and prevention systems, antivirus and anti-malware software, and encryption solutions to secure communications and stored data. In parallel, it is vital to implement a robust security policy that clearly defines roles and responsibilities, as well as procedures to follow in the event of a security incident.

The human factor is decisive in the defense against cyber threats. Workers must be regularly informed of the latest tactics used by cybercriminals and trained to recognize phishing attempts, malware and other attack vectors. Regular training sessions can greatly contribute to reducing the risk of security incidents caused by human error. Furthermore, some companies initiate and encourage personnel in risk analysis methods and suggest recommendations such as grouping access points to facilitate their monitoring.

In summary, as the importance of industrial digitalization grows, cybersecurity becomes an inseparable and crucial element of any industrial modernization strategy. Its integration from the initial stages of design or transition toward digitalized industrial systems is a requirement to protect the company's assets against the growing risks of cybercrime. This requires continuous commitment, significant investments in security technologies, active monitoring and constant awareness efforts to guarantee the safety of modern industrial environments.

2.3.3. *Environment*

The obligation for industrial companies to adopt a decarbonization strategy for production systems is more imperative than ever. As environmental concerns grow in importance, company development and growth must imperatively integrate environmentally respectful practices. The use of smart and interconnected technologies plays a crucial role in this effort, with estimates suggesting that such innovations could reduce greenhouse gas emissions by up to 15% by 2030 (WEF 2019).

The central goal of decarbonizing production systems is to minimize the environmental impact associated with industrial manufacturing. The digitalization of production systems offers several levers to achieve this. The first is the decarbonization of the manufacturing process. This includes modifying industrial processes to reduce energy consumption and, on the other hand, increase the use of renewable energies. This may also involve innovations in product design so that they are more sustainable and require less energy during manufacturing. Next come recycling and local reuse.

Indeed, modern workshops can be equipped to recycle production scrap directly on site, thus reducing the need for new raw materials and decreasing waste. After this comes energy recovery from waste: transforming non-recyclable waste into energy is a strategy to reduce final waste volumes while providing an alternative energy source. Finally, it involves maximizing the share of renewable resources in the energy mix used in order to limit the carbon footprint.

The adoption of smart industrial buildings is an example of how companies can strengthen their sustainability. These buildings can be equipped with IoT sensors that regulate energy consumption based on actual demand and usage conditions. This enables more efficient and responsive energy management, avoiding waste and optimizing resource utilization. Furthermore, smart production management systems can be programmed to synchronize energy-intensive operations with periods of increased availability of sustainable energy. This ensures that periods of high energy consumption coincide with moments when environmental impact is minimized.

In summary, the decarbonization of production systems is not only an environmental necessity but also a strategic opportunity for companies to position themselves as leaders in sustainability, while optimizing their processes and reducing their operational costs in the long term.

2.3.4. *Human dimension*

This functionality addresses the place of humans in the digital transformation of industrial processes. But this question appears emblematic of another, broader one, which addresses the evolution of professions in a projection of job destruction, which will undoubtedly be greater than that of the previous revolution where automation and robotization progressively replaced repetitive human tasks. Examples abound. Will future autonomous vehicles replace taxis in urban environments? They will be more available, safer, more economical and without any social contestation: What about the evolution of the taxi driver profession? The answer to this societal question largely exceeds the framework of our contribution, whose scope will be limited to aspects of training and employee support. Moreover, autonomous vehicles are already deployed in several regions and cities in the United States and around the world, mainly in environments where driving conditions are well monitored and where extensive testing can be carried out.

The first major challenge of the coming years will therefore be to define the new skills of the various stakeholders at the level of digitalized production lines, taking into account new execution requirements in terms of modality (on-site/remote) and temporality (real-time/deferred time). For example, one of the roles of a production manager was to organize operations scheduling. With the digitalization of production lines, their role evolves toward managing production priorities in terms of time, cost and energy consumption. As for operations scheduling, it will be dynamic and entrusted to the supervision system, which will adjust operation parameters based on the analysis of upstream data. A significant evolution of the knowledge essential to the supervision of digitalized production systems is observed. The second challenge will be to develop future supervision tools, integrating immersive and intuitive dialogue interfaces.

New production lines are smart and interconnected. Their operation requires a progressive support approach for current employees in the use of new technologies. On the other hand, it is becoming urgent to consider new training pathways dedicated to skills acquisition for future employees in order to be able to interact with information systems and exploit data. Reinventing the professions of tomorrow is thus contributing to the success of the transition toward the industry of the future while respecting societal and environmental values.

2.3.5. *Data*

The value of data for digitalized industrial processes no longer needs to be demonstrated. The analysis of collected data is today an imperative for decision-making, supervision or improvement processes. Moreover, data runs through nearly all the functionalities mentioned previously. Furthermore, data generated or used by industrialization processes positions itself within the digital flow that extends from design phases to product end-of-life management.

In the context of Industry 4.0, collected data also enables real-time monitoring of production progress. This analysis can reveal inefficiencies, predict potential failures and provide insights for substantial improvements.

The "data" functionality defines several requirements necessary for high-performance exploitation of this resource. Indeed, to fully leverage this goldmine of information, several technological requirements must be satisfied. Thus, digitalized industrial installations must have high-performance communication networks, adequate storage capacities and sufficient computational resources, part of which is located close to manufacturing lines and the other part at the Cloud level. This infrastructure must be capable of supporting the volume, velocity and variety of industrial data. Next, Cloud resources must be integrated to enable a level of processing that cannot be performed locally. The use of intelligent algorithms to analyze data makes it possible to define complex models and make precise predictions that can improve factory operation. Furthermore, it is necessary to archive production data to enable deferred-time analyses that can reveal interesting phenomenological behaviors. Similarly, data traceability makes it possible to guarantee the authenticity and verifiability of information used in decisions. Finally, data must be easily accessible to authorized stakeholders, including operators, supervisors and sometimes even customers and suppliers. Effective data sharing improves collaboration, transparency and efficiency throughout the entire value chain.

The effective adoption of these data valorization practices not only transforms production but also influences the company's position in the market. Data makes it possible to respond more quickly to changes in demand, optimize operations to reduce the cost/energy/time triad, improve product quality and minimize risks and time-to-market. In summary, structured management of industrial data is essential for any company wishing to derive added value from it.

2.3.6. *Scalability*

The concepts of adaptability and scalability mean that systems must be designed in such a way as to be able to adapt to technological changes, regulatory changes and new market requirements without requiring a complete overhaul. Here are some elements that contribute to this "adaptive and scalable" functionality.

Designing systems with an initial capacity greater than that required in the short term makes it possible to accommodate future increases in demand without compromising performance. This approach of initial "oversizing" is a prudent strategy that avoids high upgrade costs or unplanned expansion. Furthermore, monitoring the variability of system performance can identify needs for improvement or updating. Similarly, collecting and analyzing user feedback can provide insights impacting the future evolution of the system.

To this end, a modular production system enables the addition, removal or modification of modular components without disrupting the entire system. This offers maximum flexibility for adaptation to new technologies or for expansion according to operational needs. However, the integration of new equipment, of CPS configuration, requires updates at the level of IT networks, security, software, the digital twin, the MES, as well as other functional upgrades and bug fixes.

Similarly, the adoption of open standards for CPS interfacing enables greater compatibility between different components and facilitates the integration of new technologies developed by different suppliers. This also reduces dependence on proprietary solutions that may limit future system evolution options. Furthermore, the CPS must enable easy and flexible programming, with automation capabilities that can be updated or modified according to needs without requiring extensive technical intervention. The implementation of gateway applications that support user-modifiable scripts or configurations can contribute to this flexibility.

For production systems to remain adaptive and scalable, it is essential that operational personnel be regularly trained on new technologies and system updates. Investing in short-term training helps employees make technological advances more profitable and adapt to new execution modalities, which can sometimes be complex and poorly understood without prior learning.

By integrating these principles from the design or transition phase, industrial systems can not only meet current requirements but also evolve economically. This ensures that the technological investments made today remain relevant and productive over the long term, thus optimizing return on investment for companies.

3

Fundamental Concepts

To understand the digitalized industry, it is essential to master certain fundamental concepts necessary for modeling an integrated production chain equipped with advanced functionalities. The first concept concerns the data lake. The second explores in detail cyber-physical manufacturing systems from four angles: design, deployment, complexity and investment. The third addresses the Industrial Internet of Things and its impact on dynamic production supervision. The fourth concept presents industrial cyber-physical systems, which complement the design of integrated production lines to ensure a continuous flow of operations. The fifth concept deals with dynamic supervision in relation to a distributed network architecture typology Lightweight Context-Aware IoT Service Architecture (LISA). Finally, the last focuses on intelligent computing, essential for data analysis and decision-making.

3.1. Data lake

It is certain that Industry 4.0 will generate more and more data, which will constitute as much information that must be collected, analyzed and exploited to derive added value. To do this, companies will need to be capable of identifying and locating data in their original format within a data lake. The challenges of data mastery are not negligible. Indeed, according to the Data Literacy Index, the expected benefits for companies that succeed in improving their ability to use data could see their value increase by 3–5%. Furthermore, there appears to be a positive correlation between, on the one hand, improving staff capacity to use data for decision-making and, on the other hand, company performance measures such as gross margin, return on assets, return on equity and sales profitability. Hence, the importance of integrating this issue into the upstream phases of the digitalized factory design transition project. The objective is to define a process for taking into account data identification and localization aspects within a data lake.

A data lake designates a global storage space that absorbs the data flow from digitalized industrial infrastructures, such as Internet of Things (IoT) sensors, Supervisory Control and Data Acquisition (SCADA) systems, computer numerical control (CNC) machines and industrial vision devices. This type of storage must enable great flexibility to interact with data in their original format, whether raw, semi-structured or structured. This includes various formats such as CSV files, JSON, machine logs, images, videos or even real-time data generated by sensors. For example, vibratory and acoustic data collected at high frequency from sensors mounted on a rotating machinery, such as turbines or electric motors, can generate enormous volumes of data. This data, often recorded continuously for predictive maintenance analyses, is storage-space intensive, requiring efficient archiving strategies to manage terabytes, or even petabytes, of information per year.

Beyond simple storage, one of the major challenges of the data lake is to facilitate advanced computational processing to extract actionable information, not only for the optimization of production processes but also for design phases, predictive maintenance and continuous product improvement. This processing includes the use of machine learning algorithms, statistical analyses and numerical simulations to extract decision models that can directly impact operations scheduling and production forecasting. Operations scheduling, which consists of planning and organizing the sequencing of production sequences, can be optimized through the analysis of data from the lake, making it possible to minimize downtime, avoid bottlenecks and adjust priorities according to real-time constraints, such as machine availability, demand variability or unexpected breakdowns.

The architecture of a data lake must be designed to ingest, store, process, secure and analyze large quantities of data from various sources, particularly in the context of digitalized industrial systems. A data lake architecture includes several layers and pipelines that ensure a fluid and secure data flow (see Figure 3.1). The first component of this architecture is data collection and transfer. Data sources may include databases, IoT systems, production plans, maintenance plans, business applications and application programming interfaces (APIs).

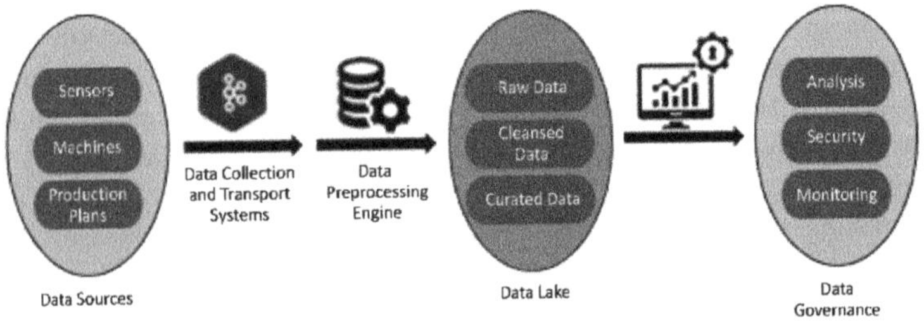

Figure 3.1. *The cleansed data pipeline*

Next, data is stored in different zones. The raw zone stores data in their native, unmodified form, thus preserving the integrity of original data for future reanalyses. The cleansed zone contains data that have been cleaned and validated. The curated zone is reserved for data ready for analysis. This process includes data enrichment, normalization and aggregation, thus making them exploitable for more advanced analyses. Furthermore, security measures are implemented, including access controls, encryption of data at rest and in transit, and regular security audits to ensure compliance and protection of sensitive data.

Data analysis is facilitated by the integration of various machine learning (ML) tools and advanced analytics. In addition, APIs and access services expose the data, allowing other applications to analyze them directly. Finally, monitoring and observability are integrated to ensure the proper functioning of the entire architecture. Performance monitoring tools track the status of data pipelines, while alert and notification systems are established to detect and report performance, security or data integrity issues, thus ensuring a robust and reliable data lake architecture.

The choice to locate data processing elements close to their sources or in a Cloud depends on the intended use. For example, video streams from industrial surveillance cameras or vision systems for quality control can be analyzed by artificial intelligences deployed locally, such as deep learning models embedded on dedicated graphics processing units (GPUs), to enable rapid anomaly detection on a production line. This approach, called edge computing, is particularly useful for applications requiring low latency, such as real-time process control and the automation of immediate corrective responses.

Conversely, more complex analyses that require greater computing capacities or comparisons of historical data are often performed on Cloud infrastructures. For example, the analysis of vibratory data generated by sensors mounted on rotating machinery, combined with dimensional and geometric measurements of finished parts, can reveal correlations between equipment performance deviations and observed drifts in product quality. The use of predictive models based on ML in the Cloud makes it possible to identify trends, prevent potential breakdowns and optimize maintenance schedules, thus improving equipment availability and reliability.

Thus, the data lake becomes an essential component of Industry 4.0, supporting not only massive data storage but also their intelligent exploitation to improve decision-making, parameterize operations, optimize flows and reduce the environmental footprint of production systems.

3.2. Cyber-physical manufacturing system

Cyber-physical manufacturing systems (CPMS) embody the future of digitalized industrial processes, marking a significant advance toward smart and interconnected factories. Unlike traditional manufacturing systems, CPMS deeply integrate physical components (such as machines, robots, sensors and actuators) with cybernetic systems (management software, computing tools and intelligent algorithms). This integration creates a dynamic and flexible network where each physical element is connected to a digital replica, often called a "digital twin", which monitors its operation and performance in real time. Thanks to embedded intelligent algorithms, CPMS are capable of collecting, processing and analyzing local data to identify various situations, both concerning the operation being executed and the state of the machine or industrial equipment. They then transmit this information to the MES, thus enabling optimized supervision of production flows.

One of the main assets of CPMS lies in their ability to establish continuous bidirectional communication between physical processes and digital management systems. Data collected via sensors integrated natively into CPMS or added for specific needs are preprocessed and stored in the data lake. Depending on the supervision architecture, this data can be analyzed and support operational decisions made in real time by the manufacturing execution system (MES). For example, in the event of detecting an anomaly or potential malfunction, the system can reorganize operations scheduling autonomously, trigger corrective actions or reconfigure equipment without human intervention. This advanced interaction between data analysis, real-time decision-making and the characteristics of CPMS confers upon production systems great responsiveness, making production lines more agile and resilient in the face of operational disruptions or evolving customer needs.

CPMS also leverage cutting-edge technologies such as cloud computing, artificial intelligence (AI) and edge computing. Cloud computing makes it possible to centralize and analyze data on a large scale, revealing complex phenomenological behaviors and facilitating the development of predictive or monitoring models dedicated to the nature of the processes that characterize a given CPMS. This capability enables active monitoring of operations status, alerts in the event of deviations and optimization of process parameters to improve operational performance. For its part, edge computing analyzes data as close as possible to their sources, an essential asset for applications requiring rapid decision-making, particularly quality control.

The flexibility of CPMS is also manifested in their ability to be reconfigured. Unlike traditional systems, often rigid and difficult to modify, CPMS enable rapid adaptation to production changes, such as the addition or removal of physical components or the introduction of new technologies. This modularity is made possible thanks to an open and interoperable architecture, which allows smoother integration or removal of the various components of the physical system as well as its virtual twin. This modular approach contributes to reducing upgrade costs and increasing operational capacity. However, to maintain the coherence of the digitalized production system, it is essential to ensure updating of the global digital twin responsible for supervising the production line. Furthermore, this modularity sometimes encounters resistance from certain manufacturers, who express security concerns regarding the complete opening of CPMS, which greatly limits the expected flexibility.

CPMS occupy a central place in the realization of the factory of the future, where systems are not only intelligent but also collaborative. The various entities of the CPMS (machines, robots, sensors) interact autonomously, exchange information and make decisions within a local perimeter. For example, in an automated production line, a robot can communicate with a machine to verify whether conditions are met before loading it with raw parts. Similarly, optical scanners measure and evaluate in real time the quality of parts, and transmit this data to the CPMS, thus enabling immediate adjustments to maintain a constant quality level. This level of integration, which promotes local exchanges between nearby industrial equipment, not only improves efficiency and responsiveness to variations but also reduces the burden on communication networks by avoiding centralizing all decisions at the level of the MES, which instead focuses on a global functional vision.

The impact of CPMS also extends beyond the factory walls. Their connection via external networks to manufacturers' remote monitoring systems, particularly for complex CPMS, enables active and continuous monitoring of their behavior, thus anticipating any performance degradation, which increases their availability and reliability. This seamless interconnectivity creates more coherent industrial ecosystems but simultaneously raises cybersecurity questions. CPMS contribute to building a global, agile and resilient production network, capable of responding quickly to disruptions and leveraging innovative functionalities.

In summary, CPMS are revolutionizing the industrial landscape by offering solutions for smarter, more connected and more flexible factories. By integrating advanced technologies in data capture, intelligent computing and hybrid communication networks, CPMS enable industries to strengthen their competitiveness. However, this transformation requires careful management due to increasing technological complexity and specialized skill requirements. The development of CPMS marks a crucial step toward an industry where processes are not only optimized in real time but also capable of adapting to an evolving environment.

3.3. Industrial Internet of Things

All experts affirm that the Industrial Internet of Things will impact the economy of tomorrow. By 2030, it is globally forecast that 40.1 billion objects will be connected to the Internet, an expansion driven by sectors such as industrial automation, healthcare and smart cities (ABI Research 2024). This technology obliges companies to consider the advantages they can derive from it, particularly through access to new business models that extend to the valorization of services associated with marketed products.

The Industrial Internet of Things (IIoT) is defined as the use of IoT technologies in the industrial sector. IIoT integrates big data, AI, as well as machine-to-machine (M2M) communication, and consists of enabling two or more machines to communicate without human intervention. The use of IIoT is not limited to the industrial or agricultural sector; it extends to local community management, particularly for public lighting management. IIoT devices, composed of intelligent sensors/actuators, networks and software (agents), can monitor, collect, exchange, analyze and act instantly on CPMS to change their behavior or environment, all autonomously. Indeed, some authors consider IIoT devices as components of cyber-physical manufacturing systems. Obviously, definitions are scalable, marked by the viewpoint of those who employ them.

On an operational level, the Industrial Internet of Things, still little exploited to date, presents numerous advantages. This technology makes it possible to connect all the actors of the value chain – products, processes, collaborators, suppliers, customers, infrastructures and energy management – generating a considerable quantity of varied data. From the point of view of production system supervision, the analysis of generated data contributes to increasing the usability of CPMS, maximizing their availability, reducing their energy consumption and improving decision-making. Similarly, the supply chain can also benefit from IIoT technologies, particularly in configurations where IIoT systems are interconnected. Indeed, thanks to the presence of sensors at workshop inventory levels, the traceability of parts being manufactured, real-time data on supplier deliveries and information on transport conditions, Enterprise resource planning (ERP) systems can activate targeted replenishment orders, while optimizing the necessary logistics.

However, the implementation of IIoT systems is not without difficulty for companies; it conceals an intrinsic complexity. Due to the extension of the IT architecture, interoperability requirements and associated changes, the deployment of IIoT projects becomes laborious. Research by McKinsey and the University of Oxford has shown that, on average, large IT projects exceed budget by 45% and time by 7%, while delivering 56% less value than anticipated (Flyvbjerg et al. 2022).

Furthermore, IT security is a crucial element to consider upstream of the design and deployment of an ultra-connected and communicative system.

IIoT systems will extend existing IT infrastructures by adding sensing, transmission and action capabilities in production, logistics and energy management. This integration entails increased complexity and risks during implementation, particularly in terms of security, data management and maintenance of interconnected systems. Moreover, to fully exploit the advantages of IIoT systems, it is essential to integrate advanced applications, some of which operate continuously, for predictive analysis, performance measurement and real-time decision-making, which can increase interoperability risks due to system fragmentation. Finally, the management of IoT systems presents challenges related to calibration, updating, technological evolution and monitoring of system operability.

Furthermore, the IIoT system installed in a workshop must be capable of integrating a larger ecosystem comprising suppliers, customers and the supply or distribution chain, with access to data regulated by an access control scheme. This ecosystem can also develop to include functionalities for tracking certain influential external indicators, such as material prices, weather forecasts, retail sales, etc.

It is interesting to briefly discuss the different stages of implementing an IoT system. The design of an IoT system in a workshop begins with the identification of specific needs, such as performance objectives, the types of data to collect and the processes to monitor. Once these needs are defined, the system architecture is developed, integrating sensors, communication devices and data processing platforms. The selection of components follows, with the choice of sensors adapted to workshop conditions, communication gateways and applications for data collection and analysis. During installation and configuration, sensors are mounted and configured to ensure accurate data collection and fluid communication. Finally, integration with existing systems, such as MES or ERP systems, is carried out to ensure effective synchronization of data and processes, requiring a testing phase to guarantee compatibility and overall proper functioning of the system. The results of the analyzed data can then be integrated into the digital twin in the form of graphical indicators, allowing the various actors to track and assess performance evolution in real time.

This raises the question of analyzing and interpreting collected data to extract useful "business" knowledge for the production process. To this end, AI plays a crucial role. For example, ML is used to identify patterns in complex data and understand various situations, such as the machine door not being closed. Deep learning (DL) makes it possible to analyze large quantities of complex data, such as images or signals, to detect anomalies. Advanced statistical analyses predict future events based on historical data, while natural language processing (NLP) extracts relevant information from unstructured documents, such as operator comments on machines, in order to identify recurring problems and suggest improvements. The transformation of data into "business" information relies on models that learn from raw data and identify patterns useful at each stage of their lifecycle, whether it be their creation, storage or subsequent exploitation.

IIoT systems actively contribute to making the execution of production plans more dynamic. Traditionally, planning relies on models that presuppose a stable environment and monotone production. These conventional models do not take into account system dynamics, such as changes in resource states or breakdowns, nor environmental dynamics, such as changes in demand, volumes and certain production priorities. To solve this problem, a global approach combining IIoT concepts, multi-agent systems (MAS) and intelligent MES can drive real-time operations supervision, promoting adaptation to unforeseen events and increasing the flexibility of production lines. However, the superimposition of different multi-agent/IIoT/MES platforms accentuates the need for interoperability to enable smoother data exchange.

In conclusion, the IIoT is revolutionizing the industrial sector by enabling real-time collection and analysis of data from machines, products and processes. This technology optimizes production by improving inventory management, parts traceability and the supply chain, and it contributes to the development of new functionalities such as predictive maintenance and increased flexibility through artificial intelligence tools. Although IIoT contributes to dynamic operations supervision and better adaptation to unforeseen events, its implementation presents complex challenges, particularly in terms of system interoperability, data security and information management. To fully leverage the advantages of IIoT, it is crucial to adopt integrated approaches that overcome these challenges and ensure effective synchronization of data and processes within industrial infrastructures.

3.4. Industrial cyber-physical system

Industry 4.0 enables the manufacturing of diversified products on the same production line, ranging from high-value-added customized products to low-margin conventional products. To successfully transition toward the digitalization of manufacturing processes, it is essential to consider the various necessary industrial cyber-physical systems, which complement cyber-physical manufacturing systems. Unlike the latter, which focus solely on manufacturing processes, industrial cyber-physical systems encompass the complementary industrial installations, equipment, and materials necessary for the smooth and autonomous functioning of digitalized production lines. Their principles do not differ from those of cyber-physical manufacturing systems mentioned in section 3.2.

In this context, industrial cyber-physical systems are classified mainly into two categories. The first includes means of transferring components between the various workstations for manufacturing, storage, quality control and assembly (see Figure 3.2). Collaborative robots and autonomous guided vehicles are important tools for improving production flows and maintaining regularity in operations. The second category includes equipment complementary to 4.0 production systems, such as integrated quality control devices (optical scanners, cameras) and additive manufacturing technologies for producing wear parts, such as robot gripper fingers, adapted to product specificities. Additionally, augmented reality tools can be used to visualize machine status or to assist maintenance operators.

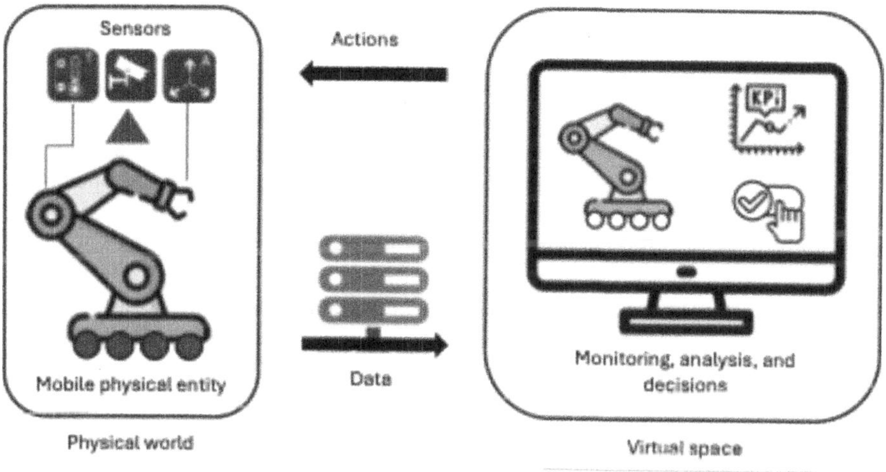

Figure 3.2. *The industrial cyber-physical system*

The acquisition of these systems represents a significant investment, both for the purchase of equipment and for staff training to develop the necessary skills. Deployment challenges are numerous, including technical aspects such as interoperability, security, training and maintenance, as well as strategic considerations such as dependence on advanced technology suppliers.

New business models are emerging, offering companies the possibility to choose between purchasing or leasing equipment, with usage-based remuneration, in order to better meet flexibility needs and cost optimization.

3.5. Dynamic supervision

Dynamic supervision, in the context of real-time operations planning and orchestration, is an approach that makes it possible to continuously adjust production processes by taking into account current conditions, optimization priorities and unforeseen events. Unlike static planning methods, where decisions are based on forecasts and largely fixed plans, dynamic supervision offers increased flexibility by adapting operations as circumstances change.

This is possible thanks to the integration of real-time data, collected from sensors, IoT systems and other sources. Decisions are then made based on the analysis of this data, enabling the system to react immediately to emerging situations such as machine breakdowns or variations in production conditions. This real-time adaptability ensures that processes remain optimized and efficient, even in the face of interruptions or sudden changes.

Dynamic supervision also promotes process flexibility by enabling rapid modifications of production operations. This includes adjustments in task scheduling, resource allocation and inventory management, in order to meet specific needs or emerging priorities. The dynamic orchestration of resources and operations is often facilitated by CPS for parts transfer and intelligent technologies, ensuring a continuous flow and efficient process management.

Moreover, this approach is essential for the operation of a digitalized industrial process. The large quantity of upstream data and the number of variables to which the system must assign values or variation functions require sophisticated, intelligent supervision and powerful computational resources, particularly for multi-objective optimization algorithms called upon at the slightest change.

Due to this complexity, the role of supervisors in production management has evolved. They must now manage various situations or undesirable events arising from internal and external causes. At the same time, they adjust the proportionalities of optimization sliders for each component or product. This includes manufacturing time, energy consumed and operation costs, taking into account customer requirements, the workshop load plan and execution conditions, such as maintenance or supply.

For example, if the customer requires rapid manufacturing while minimizing energy consumed, the supervisors will adjust the sliders for manufacturing time and energy consumption to the minimum possible. Thus, bi-objective optimization algorithms activate to suggest adequate scheduling and define values for process parameters, offering a compromise between execution speed and the lowest possible energy consumption.

Thus, dynamic supervision integrates continuous process optimization. Systems use feedback to continuously refine and improve operations execution, based on past performance and current results. This approach strengthens decision-making autonomy through artificial intelligence and ML algorithms, thereby reducing the need for human intervention and increasing responsiveness as well as precision.

In summary, dynamic supervision offers a more reactive and flexible orchestration of operations, by enabling real-time adjustments based on upstream data. This approach improves operations flow, optimizes the load plan and resource allocation in the workshop, reduces downtime and adjusts scheduling according to specific priorities such as time, energy and cost.

Nevertheless, real-time dynamic scheduling conceals increased complexity at several levels. First, methodological complexity manifests itself mainly in the design phases, where it is crucial to integrate complete modeling of production processes. Next, technological complexity concerns distributed computing, which requires the coordination of several processing units distributed across different nodes to analyze large quantities of data in real time. Finally, algorithmic complexity lies in the combined application of metaheuristic resolution methods, such as genetic algorithms, and ML techniques. Metaheuristic methods efficiently explore vast and varied solution spaces, while ML models behaviors from data, thus improving decision-making and system adaptation in real time. Taking these aspects into account makes it possible to reduce risks and size the project from different angles, ensuring the creation of solutions that are both robust and high-performing.

3.6. Intelligent computing

The combined exploitation of intelligent algorithms and computing resources constitutes two essential elements for the transition toward digitalization of industrial processes, particularly in the fields of manufacturing and production. Certain analyses require continuous computing power, sometimes of several tens of teraflops, to provide near-real-time information, enabling increased responsiveness in decision-making, particularly when using continuous video streams from cameras or optical scanners for geometric or dimensional characterization. Other analyses, particularly those concerning massive archived data, require computing resources that can reach hundreds of petaflops, a capacity that only certain Clouds are capable of providing. These computations are indispensable for identifying phenomenological correlations between different events and, consequently, for developing behavioral models usable by supervision systems.

Intelligent computing methods adapted to Industry 4.0 include the use of ML algorithms, such as neural networks and decision trees, which learn and adjust according to new data to improve decision-making processes. Genetic algorithms are also employed for dynamic operations scheduling, efficiently exploring vast solution spaces, which makes it possible to adjust production parameters in real time in response to variations in operational conditions.

Furthermore, clustering algorithms, such as k-means, group similar datasets in order to identify hidden patterns or trends, thus facilitating the understanding of complex relationships in industrial systems. Finally, NLP methods are used to extract relevant information from textual data generated in these systems, such as quality control reports, often edited by software that compares scanned models with computer-aided design (CAD) models.

To illustrate the applications of intelligent computing, 3D optical measurement quality control technology provides a good example. The complex algorithms used to reconstruct surface models from a point cloud are very demanding in terms of computing resources, especially when the required accuracy reaches precisions on the order of one hundredth of a millimeter. These computations are essential for guaranteeing product quality and precision, but they require significant processing capacities so as not to disrupt the continuous flow of production.

This technology, which relies on advanced computations, improves productivity by measuring dimensional or geometric drifts throughout the execution of a production plan. Thus, the analysis of these results by AI makes it possible to make adjustments at the manufacturing operations level before drifts tend to exceed defined tolerances, which limits stoppages for non-conformity, guarantees operational continuity and ensures compliance with deadlines.

The other technology, demanding in terms of computing resources, lies in the continuous analysis of real-time video streams. The use of cameras in the manufacturing process, combined with image interpretation by algorithms based on neural networks, brings real value for operations control and dynamic supervision. For example, a visual inspection system using video coverage of the entire production line makes it possible to visually compare, in real time, the conformity of the production process to the plan defined in the MES, visible in 3D in the digital twin.

At the algorithmic level, this involves reconstructing virtual volumes from images captured by cameras, at a rate that can reach 60 images per second, and comparing them with their geometric models integrated into the digital twin. This comparison concerns several aspects: the morphology of detected objects, their positioning and the absence or presence of objects in the workspace. Depending on the available computing power, the comparison can extend to calculating the differential between the real kinematics and that of the digital twin.

This intelligent surveillance system, integrated into the MES, is capable of signaling any anomaly, such as the presence of a part in an unknown position, the detection of a person within the robot safety perimeter or even the anticipation of malfunctions such as collisions. For example, object detection algorithms, based on convolutional neural networks (CNNs), can identify anomalies with high precision in real time. Furthermore, motion tracking techniques, such as Kalman filters, make it possible to predict potential trajectories of objects and people in the workspace.

Although this system represents a significant advance toward fully digitalized production, it nevertheless requires the development of sophisticated algorithms to guarantee satisfactory accuracy. Moreover, the requirements in terms of real-time processing, with computational capacities reaching up to several teraflops, are essential to ensure system responsiveness.

Regarding the location of computing processes, two main approaches can be considered. The first is remote computing, often referred to as cloud computing. This concept involves the relocation of algorithms and computing software to the cloud, or the direct use of algorithmic and computational resources offered as services by cloud providers. This approach enables companies to leverage the latest advances in intelligent data processing and has large storage capacities. However, it raises important questions concerning data security and confidentiality, as well as the risk of service interruption. Indeed, interruptions or malfunctions in cloud services can lead to significant disruptions in production processes, thus generating penalties due to the delays caused. Furthermore, cloud services are not always recommended for high responsiveness needs because the latency inherent in data transmission to and from the cloud can impair the processing speed required for applications demanding real-time response.

The second approach is computing at the periphery of the production site, often referred to as edge computing. Unlike the relocation of data processing to an external cloud solution, edge computing processes data directly at the edge of networks, i.e. where data is generated. This method makes it possible to reduce latency, thus offering fast response times, essential for applications requiring real-time responsiveness. This is particularly relevant in certain industrial environments where rapid decisions are crucial for maintaining operational efficiency and safety. By processing data locally, edge computing avoids the delays associated with data transmission to and from the cloud.

Furthermore, this approach offers better protection of internal data. Thus, sensitive information remains within the company's local network, reducing the risk of data loss or compromise during transfer to external servers. For sensitive industrial sectors, this protection is essential for maintaining data confidentiality and integrity, which is crucial for company stability.

Furthermore, computers integrated into cyber-physical systems (CPS), when they do not serve solely as a gateway to the cloud, are considered edge computers within the framework of edge computing. These devices are capable of managing computations locally while interacting with their production environment. This enables better distribution of computing resources at the workshop level and, consequently, optimizes the performance of industrial processes by ensuring real-time data analysis and processing nearby.

At the supervision level, as mentioned in the previous chapter, intelligent computing also enables advanced automation of decision-making. Systems can automatically adjust operations according to various criteria, such as product quality, production costs and energy consumption. This makes it possible to adapt processes to current conditions and respond effectively to disruptions or internal and external variations.

4

Concepts, Models and Architectures

The terms "concepts" , "models" and "architectures" in the context of digitalized production systems can refer to quite distinct definitions, particularly in terms of management, structure, flow and organization of production operations. A "concept" presents a global vision of the key components of Industry 4.0 and their interfaces, thus facilitating an intellectual understanding of this industrial revolution. The concept helps stakeholders, notably engineers and decision-makers, to better position these technologies within a digital transformation perspective. This theoretical understanding serves as a foundation for making the principles of Industry 4.0 tangible because the Fourth Industrial Revolution marks a break from the previous one and can conceal an intrinsic intellectual complexity.

A production "model" illustrates the way in which a company configures its organization to align with the principles of Industry 4.0. This organizational configuration influences not only the company's external environment, such as its interactions with customers and suppliers, but also the operations supervision logic and internal organization, which evolve profoundly. This includes the adoption of flexible and intelligent production processes (such as on-demand manufacturing, mass personalization or pull flow production), the integration of data-based management methods (such as predictive analysis, preventive maintenance or automated continuous improvement), the use of advanced technologies (such as Artificial Intelligence (AI), the Internet of Things (IoT) and digital twins), as well as dynamic optimization of facilities, combining physical layout and automated real-time logistics management.

The term "architecture" of production systems designates, on the one hand, a structure that organizes decision-making and information flows. Traditionally, this decision-making is centralized, flowing from the top of the pyramid (management) to the base (operational employees), facilitating a clear segmentation of roles and responsibilities. On the other hand, it can designate the way in which control is structured within the company. To this end, the pyramid architecture, implemented in the 1990s, long served to structure control and production management systems in automated environments. However, with the advent of Industry 4.0, this architecture is undergoing a major transformation to meet new technological and organizational requirements, necessitating greater flexibility and integration.

Indeed, Industry 4.0 is distinguished by an innovative distributed architecture that relies on a synergy of advanced technologies to transform production environments. At the heart of this architecture, IoT connects various machines and cyber-physical systems, thus facilitating continuous data collection. These large volumes of data are managed and analyzed thanks to big data and cloud computing, offering a powerful and elastic infrastructure for large-scale data storage and processing. AI and machine learning (ML) exploit this data to optimize processes, improve predictive maintenance and automate real-time decisions. With the addition of edge computing, this architecture evolves to process data close to their source, thus reducing latency and improving system responsiveness. Together, these technologies form an interconnected ecosystem that enables smarter, more flexible and more efficient production, embodying the principles of Industry 4.0. This modern approach contrasts sharply with the traditional pyramid architecture, where communication and data processing are centralized and hierarchical, often resulting in less responsiveness to dynamic production requirements.

The transition toward distributed architectures within the framework of Industry 4.0 constitutes a real challenge for companies, which must not only rethink the way they manage production but also how they process information at all levels of the organization. This evolution requires a profound modification of existing technological infrastructures, where centralized and hierarchical systems give way to more fluid and interconnected networks that promote rapid and decentralized decision-making. The management of these advanced technologies as well as the analysis of the data flows they produce require specialized skills in data analysis, cybersecurity and AI tools; expertise that can be rare and costly.

Similarly, the transition constitutes a major challenge for research, both in terms of design, methodology, technology and algorithms. To this end, hybrid architectures have been developed and experimented with in order to steer the transition while taking into account existing systems. This strategy relies on the pyramid architecture, widely used in industry, and adapts it to the integration of

technologies specific to Industry 4.0. This approach not only enables access to new functionalities but also promotes a gradual transition toward digitalization, starting from the existing production system.

A gradual transition toward digitalized industrial systems offers several advantages that can reassure companies and encourage the adoption of Industry 4.0 technologies. By opting for a gradual transformation, companies can minimize the risks associated with major technological upheavals while capitalizing on existing equipment. Ultimately, this allows them to gain confidence in the integration of new digital solutions. Similarly, a gradual transition gives the company the necessary time to develop and adapt its employees' skills. This includes not only technical skills but also the ability to work in more collaborative and data-driven environments.

This chapter presents the general concepts characteristic of digitalized industrial systems. It then develops an architecture detailing the configuration of connected industrial systems. Finally, it proposes a global transition model, integrating the key stages for the progressive adaptation of production systems to digitalization.

4.1. General concepts

General concepts offer a global vision of the main pillars on which a digitalized industrial system should rely. This representation, more intellectual than operational, makes it possible to visualize the principal components and their interactions. To illustrate this, three representative concepts often encountered in industry will be presented: those proposed by Gorecky and Weyer (2016), Lina et al. (2019), as well as Benfriha et al. (2021). In general, although these concepts are equivalent, their specificity depends on the context of use.

4.1.1. *Lina et al.'s concept*

Yu-Ju Lina et al. published in 2019 a concept based primarily on three technologies: cyber-physical systems (CPS), big data and the Industrial Internet of Things (IIoT), as shown in Figure 4.1. The authors emphasize the difficulties of big data tools in absorbing and processing a mass of real-time data related to the operation of 4.0 production systems. This is especially the case since the data generated is raw, unstructured, disorderly and comes from varied sources, which leads to the need for NoSQL databases.

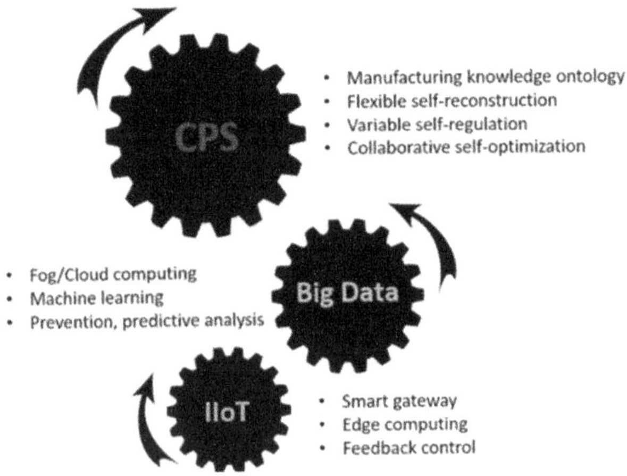

Figure 4.1. *Lina et al.'s concept (2019)*

Furthermore, Lina et al. (2019) compare the advantages and disadvantages in terms of data processing according to "deferred" and "real-time" modes. Thus, for deferred mode, cloud computing is more suitable, providing the infrastructure and services necessary for industry. There are three categories of services: Infrastructure as a Service (IaaS), Platform as a Service (PaaS) and Software as a Service (SaaS). On the other hand, network latency and data processing delays are not compatible with real-time responsiveness needs, hence the interest of edge computing, which offers several advantages compared to cloud computing, such as low latency, better quality of service (QoS) and increased data security. The authors also mention different technologies, notably the integration of multi-agent systems (MAS) under a service-oriented architecture (SOA), which enables manufacturing systems to increase their performance in terms of autonomy and intelligence. In summary, this concept is based on technologies rather than functionalities, as proposed in section 4.1.2.

4.1.2. *Gorecky and Weyer's concept*

Gorecky and Weyer published in 2016, in a SmartFactoryKL white paper, a concept for 4.0 production factories, adapted for discrete flow systems. The latter serves as a framework for the design and installation of modern production systems, characterized by operational flexibility, individualized production, and internal and external networking. It is complemented by a set of technical specifications, related to information technology, developed and applied to the SmartFactoryKL pilot factory.

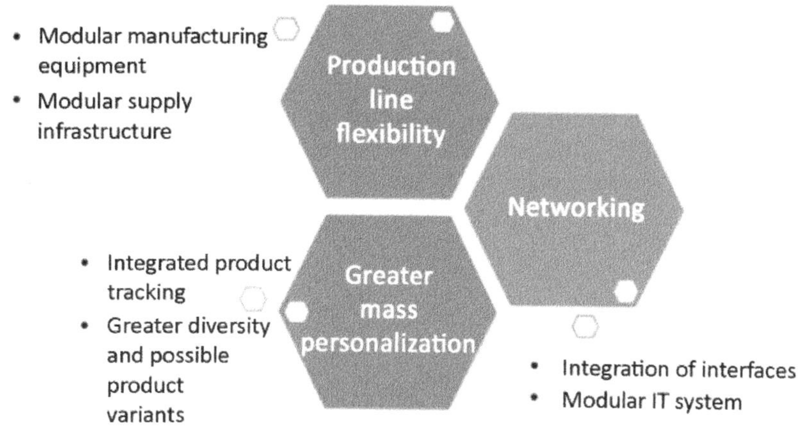

Figure 4.2. *Gorecky and Weyer's concept (2016)*

In summary, Gorecky and Weyer (2016), authors of this study, built their concept on 3 pillars, as shown in Figure 4.2. The first considers that the production factory is flexible and structured in a modular manner. Production modules can thus be reconfigured and commissioned quickly at lower cost. Moreover, for operational efficiency, this factory must benefit from a modular supply infrastructure. The second pillar introduces the factory's capacity to manufacture several variants of individualized products, which implies batch production, of modest size, under mass production conditions. To this end, the operations carried out in the production module must be designed to handle a wide diversity of products and variants. The third pillar consists of networking products and production lines by effortlessly connecting them to the company's IT systems, which implies an open IT architecture for loose coupling between IT systems and all the factory's industrial processes.

This approach, to be placed in the context of 2016, enables the production system to benefit from new functionalities, notably the ability to increase the flexibility of discrete flows. Furthermore, the authors introduce the idea of interconnecting remote production systems to optimize the use of resources and production means.

4.1.3. *Benfriha et al.'s concept*

Benfriha et al. introduced in 2021 a new general concept that reorganizes and integrates the various technologies to offer a unified and coherent vision of a 4.0

production system. Figure 4.3 shows the five essential pillars of this system. This graphical representation is designed to be clear and pedagogical; it distinguishes itself from previous ones by the integration of communication networks, which play a fundamental role in the design of digitalized industrial systems. Furthermore, interaction between the different pillars exists and can vary according to the observer's perspective (Benfriha et al. 2021).

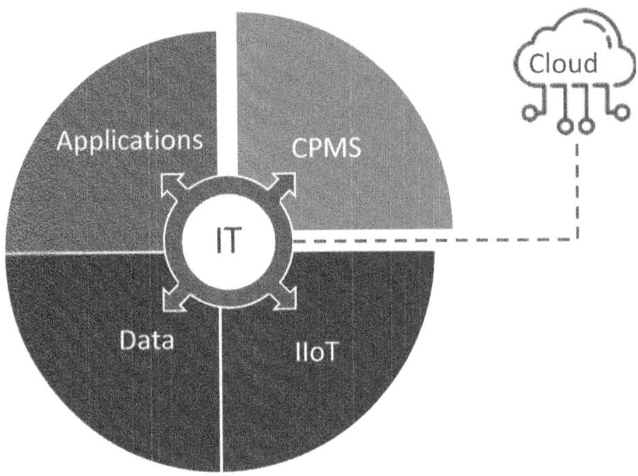

Figure 4.3. *Benfriha et al.'s concept (2021)*

The first pillar stipulates that the manufacturing workshop must be configured as a cyber-physical manufacturing system (CPMS). Machines must either be already adapted to operate in a connected environment or be capable of being upgraded, for example, by adding sensors and control cards. This upgrade is necessary to meet the bidirectional operation requirements of a 4.0 production system. This configuration makes it possible to collect data on machine status and receive control commands through the local network. Furthermore, certain machines can integrate intelligent applications, such as MAS, to analyze their own data and interact with their immediate environment.

The second pillar focuses on the integration of IIoT. This technology complements the production line by adding a wide range of sensors, cameras and associated networks. It makes it possible to realize innovative functionalities and obtain immersive representations of manufacturing processes by means of the digital twin. Furthermore, it promotes dynamic and autonomous supervision of operations. This infrastructure also facilitates the integration of cloud computing resources for the analysis of archived data and the development of predictive models or production flow optimization.

The third pillar addresses the applications necessary for the implementation and control of a digitalized production program. It emphasizes the challenges related to the interfacing and hierarchy of the software involved. Indeed, the multitude of applications and software that interact at different stages of the manufacturing process makes the question of their organization and integration crucial. This is essential for guaranteeing a fluid and bidirectional information flow between the manufacturing process and supervision systems.

The fourth pillar addresses data management in all its aspects: storage, processing, valorization, sharing and visualization. This pillar is at the heart of the current industrial revolution and represents a major challenge in terms of technicality, skills and cybersecurity. The analysis of this data makes it possible to generate valuable insights that can considerably improve the performance of the production system. By exploiting the collected data, it becomes possible to identify phenomenological correlations and deploy predictive maintenance. Furthermore, data analysis makes it possible to supervise operations scheduling in real time and optimize supply chains.

The fifth pillar, probably the most complex, concerns technological choices relating to information networks. These networks, often superimposed, are generally designed in a hybrid manner, adopting either specific schemas or object-oriented or event-driven architectures (EDA: event-driven architecture, SOA: service-oriented architecture). This makes it possible to meet the needs of internal and external entities, which interact through different protocols and execution modalities, in real time or in deferred mode. This pillar also integrates the use of distributed multi-agent systems, capable of communicating and negotiating local solutions between different cyber-physical systems.

4.1.4. *Conclusion*

In conclusion, the various concepts studied in the literature largely converge on common fundamentals. When exploring modern digitalized systems, it appears that certain themes recur regularly in recent scientific and technical literature. Notions such as networks, connectivity, intelligent agents, interoperability, artificial intelligence, portability, distributed computing, flexibility and transversal digital continuity are often highlighted. These concepts define the foundations on which digitalized industrial systems are built and evaluated.

Although these concepts share similar foundations, each is distinguished by specific characteristics adapted to operational contexts or priorities, such as system reconfiguration capability, operational flexibility or geographical distribution of production units. These particularities make each concept high-performing in its field of application, while sometimes limiting its generalization. For example, a concept designed for the automotive industry, where mass production and robotization are common, might not be entirely suitable for a small manufacturing enterprise specialized in customized products in small quantities. This distinction underscores the need to customize Industry 4.0 solutions.

The last concept overcomes this limitation by proposing a more generic framework, representative of various industrial situations. This concept, open to different Industry 4.0 functionalities, thus guarantees broad applicability and relevance in diverse industrial environments.

4.2. Reference model

As indicated previously, the term "model" in the industrial domain refers to the definition of a global configuration that facilitates the organization and control of industrial processes. Its implementation requires the adoption of standardized interfaces, protocols and methods, independently of manufacturers or solution providers. In Industry 4.0, there is a reference model, called RAMI 4.0, which was developed in Germany (Heidel et al. 2019). In the United States, an equivalent model is known as the Industrial Internet Reference Architecture (IIRA), developed in 2019 and updated in 2022 by the Industry IoT Consortium (Industry IoT Consortium 2022).

The RAMI 4.0 model, or Reference Architecture Model for Industry 4.0, is a three-dimensional representation designed to comprehensively map the Industry 4.0 space. It is distinguished by its complex structure that helps to understand and organize the interconnected components of the digitalized industry. This space is structured along three axes: the vertical axis, which comprises six layers dedicated to industrial organization; the left horizontal axis, which represents the lifecycle of products and production systems; and the right horizontal axis, which illustrates the various functionalities within companies and factories (see Figure 4.4).

Thus, in the vertical axis of the model, the first layer called the "assets" layer exposes physical objects such as parts, documents, archives and schematics, as well as human actors. Just above is the "integration" layer where physical objects are transformed and connected to the digital world. The elements of the "assets" layer are linked to the digital world through this integration layer, which simplifies information processing and serves as a gateway between the physical and digital worlds. This layer encompasses process supervision, human–machine interfaces, networks and sensors, among others. The next layer, “communication”, ensures standardized communications between the “integration” layer and the “information” layer.

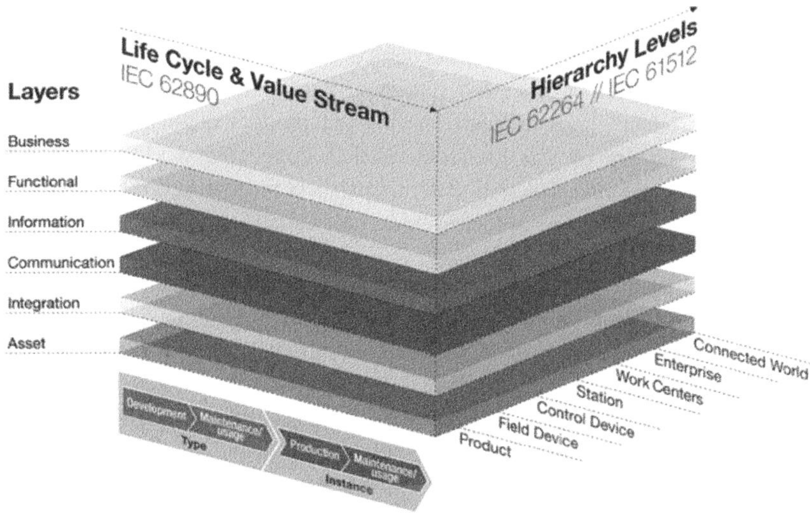

Figure 4.4. *Reference architectural model for Industry 4.0 (RAMI 4.0)*

The “information” layer, which supervises the “integration” layer, stores data in a structured manner, includes details on total sales volume, purchase orders and suppliers. It also holds data on all products and their characteristics as well as on the machines and components used in the manufacturing process. This layer also disseminates information to customers and collects their feedback. Operating on software bases, it can manifest itself in the form of applications, digital data or files. Similarly, it is responsible for converting events into data usable by the upper layers. Above it is the “functional” layer, which orchestrates production operations and system control. The last layer, “business”, defines marketing strategy, business models and commercial objectives.

The horizontal axis, on the left side of Figure 4.4, shows the lifecycle of products and production systems. It is divided into two main phases: the type (design and development) and the instance (operation and maintenance). These elements are crucial for understanding how products evolve from the initial idea to their withdrawal from the market, integrating the value flows that measure the economic inputs and outputs associated with each stage.

The second horizontal axis, located on the right side of Figure 4.4, is dedicated to the hierarchical layers of Industry 4.0. It comprises six layers: four of them, "enterprise", "work centers", "station" and "control device", are recognized by international standards, while two additional layers, "field device" and "product", are added outside these norms. The "field device" layer is dedicated to the intelligent management of machines or processes, particularly through the analysis of data from sensors. The "product" layer focuses on finished product quality and production capabilities. At the top of this hierarchy is the "connected world" layer, which ensures the factory's connection with external partners through service networks and cloud platforms, reflecting one of the essential functionalities of Industry 4.0.

The detailed analysis of the RAMI 4.0 reference model, which includes all elements of the vertical and horizontal axes, does not detail the connection modalities between the different components within each layer. These connections are essential and should be clearly defined during the design or transition toward a digitalized factory. In conclusion, although RAMI 4.0 proposes a well-structured framework, it remains relatively generic. This model helps companies explore the complexity of Industry 4.0 by offering a unified vision and precise reference points for all participants involved.

4.3. Generic architecture

Production system architectures designate a structure that organizes decision-making and information flows. The traditional architecture, already widely covered in the literature, will not be detailed here. The focus will be on a generic architecture, adapted for industrial projects in the design phase, but less for transition projects for which current production structures represent an initial constraint.

To this end, the architecture of Bonnard et al. (2019, 2021) was designed within the framework of the research project entitled "AutoManSec 4 CloudIoT" (Autonomic Management and Security for Cloud and IoT), led by Carlos Becker Westphall, from July 2017 to July 2021 (Westphall 2021). The objective was to develop and experiment with a new global generic architecture, where real-time data collection and processing constituted the central core around which the other components were integrated. This made it possible to realize new Industry 4.0 functionalities, notably data-driven operations supervision. This architecture prioritizes the collection, storage and analysis of data generated by workshops, whether by added connected objects, or by those already natively integrated into cyber-physical manufacturing systems along the production line, thus producing a considerable quantity of data. Vibratory or acoustic sensors, for example, generate a significant volume of data in a very short time.

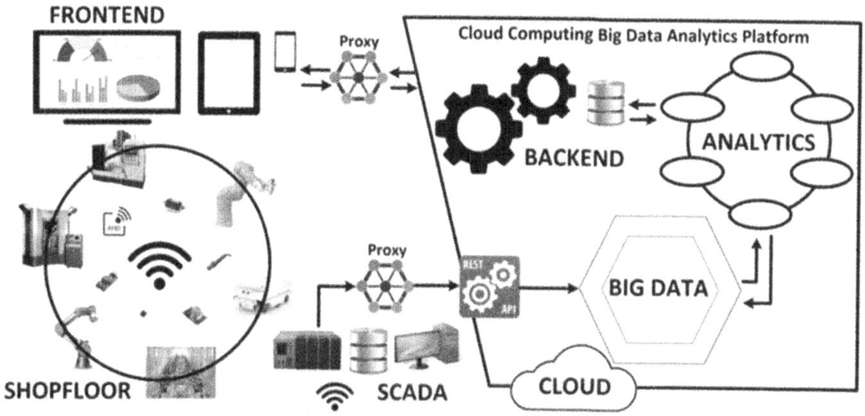

Figure 4.5. *Bonnard et al.'s archictecture (2019, 2021)*

This architecture is composed of several key elements that work in synergy to improve data management and analysis. At the heart of this architecture is a high-performance supervisory control and data acquisition (SCADA) system for monitoring workshop equipment and collecting real time data. Added to this is a monitoring system using a user interface (frontend), which enables operators to visualize operational data and react quickly to specific situations or malfunctions.

For the storage and management of large volumes of generated data, the architecture uses a cloud solution with big data technologies. Another crucial aspect of this architecture is its flexibility in terms of communication protocols. Indeed, this makes it possible to support a wide variety of data and formats, rendering the system adaptable to different types of industrial environments and sensors. This flexibility is essential for ensuring that all relevant data can be captured and used for monitoring and analyses. Similarly, the graphical supervision interface is designed to be accessible via web and mobile platforms, thus offering increased portability. Users can interact with the system wherever they are through secure access devices.

In terms of data processing, the system uses a robust backend, connected to a MySQL database. This backend embeds advanced algorithms, exploiting both traditional computing methods and artificial intelligence, in order to perform complex analyses and generate relevant operational insights. In summary, this architecture aims to reduce the complexity and costs associated with the implementation of Industry 4.0 technologies, while offering a scalable and efficient approach for data collection, storage and analysis, thus contributing to rational decision-making and continuous optimization of production processes.

4.4. Transition approaches

Before implementing Industry 4.0 technologies, it is advisable to first optimize production processes, particularly through methods such as Lean Manufacturing. This approach aims to eliminate inefficiencies and standardize operations, thus creating a solid foundation for the effective introduction of advanced technologies. Without such preparation, the integration of Industry 4.0 innovations risks not reaching its full potential.

Lean Manufacturing is essential for precisely identifying where and how Industry 4.0 technologies can be applied to maximize efficiency. It also establishes a culture of continuous improvement and highlights areas where technologies can bring significant improvements, thus ensuring an optimal return on investment. In summary, optimization through conventional approaches is a prerequisite for a successful transition to Industry 4.0. Subsequently, companies can progressively engage in a transition process, relying on a roadmap adapted in terms of priorities, constraints and investments. Generally, companies are motivated by the search for additional performance, the integration of new functionalities or by external constraints. This is particularly true when they must adapt to their ecosystems, especially when customers and suppliers request updated data on ongoing processes.

Section 4.4.1 is organized into two integration levels. The first level, which offers a conceptual description and a global vision, presents the principles of horizontal and vertical integration related to the transition of industrial production systems toward a digitalized version. The second, in continuity with the first, provides more detailed elements, characterized by a hybrid architecture for integrating Industry 4.0 technologies into an existing structure.

4.4.1. *Horizontal and vertical integration*

The principles of horizontal and vertical integration are widely recognized and approved by the majority of experts in the field of Industry 4.0. They make it possible to understand and schematize the transition toward digitalized production systems.

Vertical integration concerns the connection and communication between different levels of the production architecture within the same company, ranging from the field (machines, production equipment) up to management systems at the higher level. This includes the integration of levels such as production control, planning and enterprise management. The objective is to enable rapid and effective decision-making thanks to a continuous flow of real-time information between basic operations and strategic decision levels. For example, data collected on an assembly line can reveal in real-time deviations at the planning level, thus leading to urgent decision-making.

Horizontal integration refers to the coordination and collaboration between different production units or between different companies, sometimes across organizational boundaries. This can involve information exchange between different production sites, suppliers, distributors and customers. The objective is to optimize the entire value chain, from the supply of raw materials to the delivery of finished products to customers, by improving traceability, communication and collaboration between all stakeholders. This integration is often supported by technologies such as digital twins, which enable enhanced interaction and coordination between different actors and processes.

In summary, vertical integration focuses on improving communication and management within a company, from production to the strategic level, while horizontal integration aims to improve processes across different companies or departments to optimize the overall value chain. Both forms of integration are essential to fully benefit from the advantages of Industry 4.0, thus making companies more responsive, more efficient and more competitive in a dynamic market environment.

In the context of manufacturing production, vertical integration contributes to increasing flexibility at the level of operations, control command and data management. In contrast, horizontal integration consists of establishing fluid and real-time communication between the company's business level and production lines that may be geographically dispersed, thus enabling active and dynamic supervision of the various industrial processes and services. These approaches are compatible with the overall objective of the digital transition, which aims to improve operational performance.

Additionally, when it comes to the technical flexibility of the workshop, several aspects relating to machines must be taken into account. Machine flexibility implies addressing challenges such as interoperability, remote access and the ability to reconfigure processes flexibly. In addition, the integration of machines at a decision-making level requires their transformation into cyber-physical manufacturing systems, equipped with digital twins. The modularity of manufacturing operations also constitutes an essential factor. Thus, the overall flexibility of the production system relies on the convergence between bidirectional decision-making flexibility and operational flexibility, particularly with regard to the variation of manufacturing parameters or process reconfiguration.

Furthermore, it is essential that manufacturing and production flow supervision systems be scalable and dynamic. Once digitalized, the workshop machines must be integrated into the supervision system to become fully operational. This makes it possible to orchestrate the execution and sequencing of operations dynamically and in real time. However, the process of transforming and integrating all machines and industrial equipment into the supervision system is often time-consuming and complex. Moreover, certain prerequisites are indispensable, notably the definition of digitalized operational ranges, designed in a modular and parametric manner. This concerns operations related to processes, quality control, as well as the transfer of parts by robots or conveyors.

On the other hand, data generated locally by cyber-physical equipment, sensors or the methods department must be processed by algorithms and computing units at the edge or in the cloud. They must be classified, aggregated, transferred, stored and analyzed with a view to optimization and improvement. Consequently, the need for fluid management of data flows and adapted storage capacities requires an evolution of the IT architecture toward a distributed configuration. To this end, several solutions can be considered, such as event-driven architecture (EDA) or service-oriented architecture (SOA), often favored for manufacturing applications and software, including manufacturing execution systems (MES) and enterprise resource planning (ERP) systems. These architectures propose generic solutions, but they must therefore be adapted or combined with other solutions according to the operational context or specific needs.

Finally, horizontal integration influences vertical integration. Data, information and decisions from digitalized production systems can be exploited at the company's commercial level, particularly with regard to customers and suppliers. Natively, vertical integration makes it possible to collect real-time information on production, such as yield rate and downtime. This information can then be used to adjust sales forecasts and improve inventory planning, for example. Similarly, if a production line encounters a performance problem, the data can alert the sales team so that it can inform customers of possible delays, while optimizing supply chains by adjusting orders with suppliers. This illustrates how horizontal and vertical integration in a digital transition context can structure the transition of production systems within the company.

4.4.2. *Hybrid approach*

A hybrid architecture is an adapted version of the traditional pyramid architecture, designed to align with the requirements of a digitalization process. It takes into account existing industrial infrastructures and equipment. The objective is to offer a transition framework based on a conventional pyramid structure. This approach starts from the principle that companies cannot abandon their production facilities, which are often expensive and whose lifespan has not yet expired. The idea is therefore to progressively modernize machines, control tools and production organization toward an adapted level of digitalization, without dismantling what exists.

In this context, an initial reflection emerges to examine the evolution of a traditional pyramid architecture, operating according to a 3.0 approach, toward a distributed 4.0 architecture, as shown in Figure 4.6. This approach is of major interest to industrialists because it retains the old architecture as a base, while progressively superimposing new technologies in accordance with the targeted functionalities. Moreover, it enables a gradual transition, facilitating the appropriation of new skills and working methods. Indeed, the investment level for this type of solution remains accessible and contributes to minimizing risks.

However, this hybrid approach comes with certain disadvantages due to specific limitations. Indeed, the company that adopts this architecture might not benefit from all available functionalities nor achieve optimal performance for some of them.

For example, performance related to flexibility depends on the ability to physically and functionally integrate robots or mobile platforms, which can compromise the targeted performance.

Furthermore, some older machines operate under proprietary and closed operating systems, which complicates internal data collection and limits supervision autonomy as well as improvement initiatives. Nevertheless, this method represents a good compromise in terms of cost/benefit ratio. In most cases, a company cannot completely replace its machine fleet and does not necessarily need all the proposed functionalities. Thus, this strategy appears as a balanced solution between digitalization requirements and technological investments.

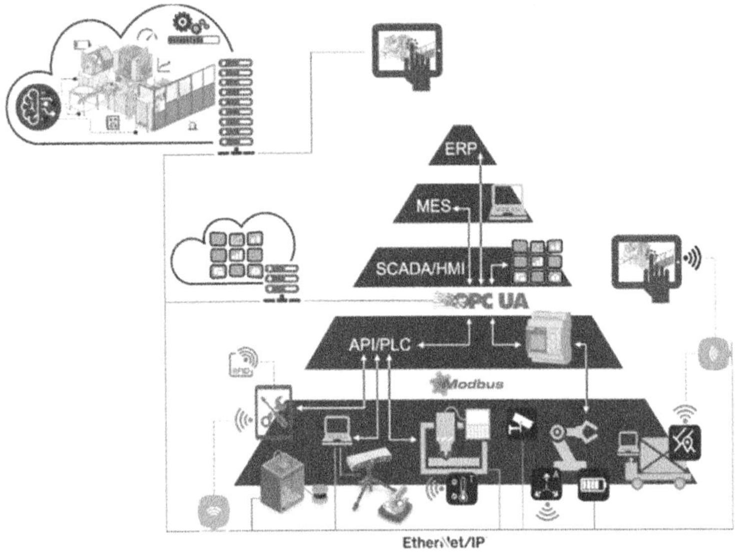

Figure 4.6. *Hybrid architecture*

The evolution toward a hybrid pyramid architecture involves five aspects. First, the hardware aspect, which includes the acquisition of complementary equipment such as mobile robots for component transfer between different machines or storage areas, contactless quality control devices and machines for additive manufacturing of supplementary components. This equipment, generally designed in a "cyber-physical" configuration, must be selected according to certain conditions in order to guarantee their compatibility and effective integration with the entire production system. It is also crucial to remain vigilant regarding associated software, particularly their ability to interact and connect with other systems. Finally, the management of incoming and outgoing data, in terms of volume, quality, availability and format, must be compatible with analysis and supervision systems.

The integration of cyber-physical production systems such as mobile robots can present additional challenges due to their increased complexity. Their operation requires the use of multiple software programs: one for the mobile platform itself, another for the manipulator devices it carries and a third for the interface with the programmable logic controller. Furthermore, the operation of these advanced technologies requires specific skills not only for their daily functioning but also for their maintenance. This implies staff training or the hiring of qualified specialists to manage these complex systems.

The second aspect involves two major components: the reconfiguration of existing machines and the integration of an IoT layer from an OT/IT convergence perspective. To effectively integrate these machines into a connected production system, it is essential to reconfigure them in terms of controls and connectivity. This often involves updating or replacing control systems so that they can communicate with supervision software and data platforms. For example, a traditional machine can be equipped with new programmable controllers compliant with the "IEC 61499" standard, which offers a more open and interoperable approach. This standard, a recent international standard, redefines the architecture of programmable logic controllers to enable them to react to events more dynamically. Its principles are based on the modularity, reusability and interoperability of functional blocks, thus facilitating the creation of flexible systems compatible with a distributed architecture.

Furthermore, the integration of an IoT layer is essential for transforming a traditional workshop into a digitalized production environment. This layer consists of sensors, cameras and dedicated networks that continuously collect data on the production process. Sensors can monitor various parameters such as temperature, pressure and speed, while cameras can be used for quality and safety monitoring. These devices must be designed to be interoperable with the company's existing networks, thus ensuring a smooth and disruption-free transition. Moreover, the direct communication of sensors and actuators with the MES through protocols such as TCP/IP, without requiring intermediaries such as programmable logic controllers, represents an additional challenge to consider in this transition. This direct approach simplifies the network structure, reduces communication latencies and improves the reliability of collected data.

The third aspect of the transition toward a hybrid pyramid architecture focuses on software acquisition, crucial for orchestrating and optimizing workshop operations. This aspect is divided into two main initiatives. The first concerns the acquisition of intelligent production supervision software. These tools enable dynamic operations scheduling, machine parameterization and workflow organization in a more autonomous manner. They also offer real-time analyses that help optimize production and reduce unplanned stoppages. MES are typically implemented to integrate all workshop operations from start to finish, thus providing complete visibility and increased control over the manufacturing process.

The second initiative consists of developing a digital environment for the industrialization phase, also called the preparatory phase for production launch. In addition to standard software, it is often essential to create complementary IT applications to meet the specific needs of the workshop. For example, this can include a production plan simulator, integrated into the digital twin. Its value lies in validating the sequence of operations before their implementation. This reduces risks, collisions and costs associated with production testing under real conditions. Furthermore, software interoperability is essential to guarantee that all components of the production system can communicate effectively, without error or delay, which is decisive when executing a complex production plan. This underscores the importance of fluid communication between software, applications and platforms, whether local or remote, used in or by the workshop.

The fourth aspect concerns computing capabilities, which play a fundamental role in the operation of advanced system functionalities. Whether deployed locally or in the cloud, they enable real-time processing of large quantities of data, by means of artificial intelligence tools, in order to provide information likely to influence production flow.

Local computing, performed directly at or near the production site, enables rapid responsiveness to operational requirements and reduces communication latency. This is particularly useful for applications requiring real-time responses, such as dynamic operations control, defect detection by vision or parametric machine adjustments. Having local processing capabilities also makes it possible to maintain operational continuity even in the event of internet connection disruptions or other remote communications. Furthermore, the use of the cloud for data processing offers algorithmic resources and practically unlimited scale for data storage and analysis. For example, advanced predictive analyses and complex scenario modeling can be carried out in the cloud, using intensive computing resources available on demand. This allows companies to benefit from powerful analytical capabilities without having to invest heavily in on-site physical infrastructures.

In this context, one of the specific challenges of Industry 4.0 production systems is the management and analysis of real-time video streams from cameras used to measure and control manufacturing processes. For example, in quality processes where each part must be scanned, processing capabilities must be sufficiently robust to analyze the captured point cloud, reconstruct the 3D geometric model and compare it to its native CAD model in order to identify defects and report anomalies within a short time frame. This is essential to prevent any deviation and maintain the required quality level without compromising production pace. By increasing computing capabilities, particularly through cloud solutions, it also becomes imperative to strengthen data security and information flow management. Systems must be designed to protect sensitive data and ensure compliance with data privacy regulations, such as the GDPR in Europe.

The fifth essential aspect of this transition is machine programming. In 3.0-type processes, machine programming is often static, designed for mass production operations where changes are almost impossible without production interruption. Each machine operates according to a predetermined scheme to maximize efficiency in repetitive productions. In contrast, 4.0-type processes adopt an approach where each machine is configured as a cyber-physical system. This allows them to modify their program based on intelligent analysis of inputs, optimization priorities or other external factors. In this perspective, it becomes essential to develop manufacturing programs whose parameters can be adapted in real time, without interrupting production. For example, robot speed can be defined as a variable that the supervision system could modify. Thereupon, the latter can recalculate and assign a new speed to the robot, thus optimizing energy consumption if this is identified as an optimization priority.

Similarly, cyber-physical manufacturing or production systems can adjust their parameters if data signals a deviation in terms of geometric or dimensional quality, without requiring human intervention or prolonged stoppages for reprogramming. Finally, thanks to the integration of AI and machine learning, machines can learn from their past performance and continuously optimize their operations while respecting the priorities set by supervisors. They can also predict potential anomalies or failures and initiate corrective actions, thus increasing the durability and availability of production tools.

In conclusion, the transition toward Industry 4.0 production systems explores different concepts and approaches to identify those that will be most compatible with the context, constraints and needs of companies, expressed in terms of new functionalities to be achieved. These concepts and approaches, which will serve as the basis for the development of a dedicated architecture, interact with various domains and areas of expertise, posing significant challenges in terms of cybersecurity, interoperability, supervision and skills. The majority of analyzed works and publications highlight a multiscale complexity that goes beyond simple methodological and technological questions. They also reveal the consequences, often perceived as negative, of the economic strategies adopted by the main players of digital platforms and Industry 4.0 technology manufacturers. These strategies include, among others, the protection of their equipment by limiting access to essential data, the promotion of business models based on usage and solution hosting, and sometimes, the insistence on remote access to data from delivered machines. This creates a dependency of companies on their industrial equipment suppliers, thus accentuating their influence and control.

5

Platform 4.0

Industry 4.0 is a young discipline that could be named "industrial digitalization science". As previously indicated, this industrial revolution is characterized by the convergence of several disciplines, notably industrial processes, automation, automatic control, mechatronics, computer science, robotics, artificial intelligence and virtual reality. It also encompasses the evolution of professions and takes into account societal and environmental aspects. This movement profoundly transforms production modes by integrating new requirements such as large-scale product personalization, decarbonization of industrial processes, use of renewable energies, recyclability of generated waste or rejects, reparability or improvement of products throughout their useful lifecycle, etc.

Many other scientific fields can be integrated into this revolution, thus amplifying its impact. For example, the evolution of trades and jobs is a crucial question for which answers are still limited, particularly due to the rapid evolution of knowledge required in an increasingly complex environment. The integration of artificial intelligence into learning new skills and into advanced process automation could radically transform the notion of employment. Jobs could thus evolve toward flexible service activities that companies would call upon according to their needs in production, maintenance, analysis or for situations that technology cannot resolve. Furthermore, the position of humans in this context is also essential, raising questions about the human–machine interaction, work modalities and employee well-being. Finally, Industry 4.0 aims to create more sustainable and responsible production systems, by taking into account their impact on communities and the environment. However, this objective must align with the requirement of profitability to encourage investments in advanced technologies.

Research in Industry 4.0 relies on both theoretical and experimental approaches, particularly to validate and evaluate the effectiveness of proposed solutions. With this in mind, the École Nationale Supérieure d'Arts et Métiers (ENSAM) launched in 2015 an ambitious project aimed at developing a full-scale production line. Since then, the total cumulative amount of this investment approaches 1 million euros from companies and institutions. This project aims to conduct research work in partnership, in order to generate new knowledge in this field and to strengthen engineering training programs with new skills adapted to the needs of the transition toward Industry 4.0.

5.1. Design and development

In its initial phase, the platform was designed and installed using technologies prior to 2015, classified as 3.0 technologies. These technologies had undergone significant improvements, thus increasing their performance. However, during this period, 4.0 technologies were emerging and were characterized by their diversity, their lack of maturity and their low compatibility with existing systems. Solutions were often fragmented, making integration with existing infrastructures difficult.

The economic challenges related to the adoption of these new technologies represented a major obstacle. Large companies hesitated to fully commit to this technological transition, due to the substantial investments required and the profound modifications necessary in their operational processes. This caution was accentuated by the lack of established standards and uncertainty regarding return on investment.

Another factor that contributed to slowing the integration of 4.0 technologies lay in the preemption of new digital market shares in industry, and this between machine manufacturers and digital companies. Indeed, machine manufacturers refused to make the "control" layer of their equipment accessible, essential for data access, for fear that digital sector companies would exploit this data through intelligent optimization applications. This would have potentially reduced manufacturers' control over their products and services, thus threatening their traditional business models. At the same time, digital companies had perceived the strategic importance of industrial data and were considering offering innovative services to machine users. This trend often came into conflict with the interests of machine manufacturers. The latter rightly considered that the valorization of data generated by their machines belonged to them as a priority. They viewed this data as an essential asset to develop and offer new services.

Thus, the initial phase of the platform was marked by the prudent use of available technologies, under constraints of divergent economic interests and data accessibility. These elements were essential for defining a platform acquisition and development strategy, which consisted of two stages.

The first stage consisted of designing and acquiring a relatively conventional production line, based on the principles of 3.0 technologies. This production line manufactured, assembled and delivered products in small series without any variation in the process, apart from the number of products to be manufactured. However, certain advanced technologies were already integrated into the initial version, such as robot supervision by cameras. Indeed, this robot–camera device makes it possible to identify and locate raw parts in the initial storage space and adapt the trajectory to pick them up. This device, natively integrated, is high-performing but not very open to the outside. Production plan control is ensured by a human–machine interface (HMI), through which operators can initialize the platform and launch the manufacturing process. This interface also enables visualization of certain information available at the programmable logic controller's register table level.

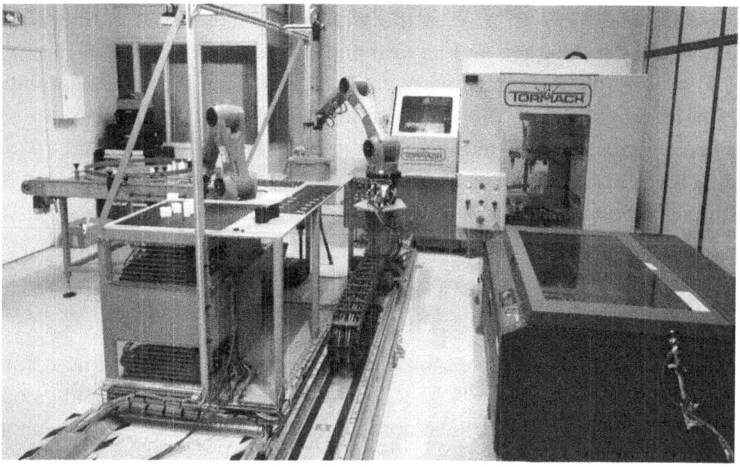

Figure 5.1. *Industry 4.0 platform in its initial version. For a color version of this figure, see www.iste.co.uk/benfriha/industrialprocesses.zip*

From the equipment standpoint, as shown in Figure 5.1, the line is installed in Zone 1 of the Industry 4.0 platform. In its initial configuration, it consists of three CNC machines, two Kuka robots (one of which is mounted on a 3-meter rail), and two cameras, dedicated to the exclusive use of the robots. Two storage spaces are also present: the first, called "initial store", is intended for raw parts, while the second, called "transition store", is used for loading and unloading the three computer numerical control (CNC) machines. Finally, a conveyor transports parts to the accumulation table, thus marking the end of the manufacturing process. Supervision is entrusted to the programmable logic controller that the operator can parameterize through its HMI. These pieces of equipment are detailed in the following sections.

The second stage consisted of inscribing the evolution of this platform within a transition approach toward Industry 4.0, thanks to an ambitious research program. This phase began in 2018 with the hosting of the first doctoral thesis work. The central theme is the digitalization of industrial production systems. The initial configuration of the platform offers an opportunity to develop innovative solutions by relying on existing infrastructures. This corresponds to the expectations of industrial partners seeking to undertake a gradual transition while taking into account production tools already in place.

The challenges of the Fourth Industrial Revolution are multidimensional, encompassing economic, social, ethical, political, cultural, educational and environmental aspects. From a scientific standpoint, they lie at the intersection of technological, methodological and algorithmic domains, developed in the following sections. However, the digitalization of industrial systems represents an ambivalent advance: it leverages advanced automation and artificial intelligence to improve performance, while risking the accentuation of social inequalities. A holistic approach is therefore essential to frame this complex transition, maximizing benefits while minimizing risks.

5.1.1. *Scientific challenges*

Each industrial revolution poses unique challenges to the scientific community. With the emergence of Industry 4.0, and by extension Industry 5.0, the singularity lies in the convergence of various domains such as technologies, methods and algorithms.

Methodological challenges focus on the development of methods that organize and structure the digital transformation of industrial production systems. Among the numerous published studies, those of Chawki El Zant (2021) stand out for proposing a conceptual transition model adapted to the traditional architecture of conventional production systems, thus promoting interoperability and operational flexibility. Similarly, Nooshin Godsian (2023) introduced an innovative approach to effectively integrate mobile robots into existing digital production lines. This technology, recognized for its complexity, requires advanced technological skills and knowledge. Finally, Nourhan Halawi-Ghoson (2025) developed a dynamic methodological approach that integrates the notions of risks and cyberattacks from the design phase or during transitions toward digitalized industrial systems. It is unnecessary to recall that cybersecurity is a priority and often an obstacle to digitalization for many companies. Recent examples have shown that the consequences of a cyberattack can even affect investor confidence.

Technological challenges in the context of Industry 4.0 involve the creation of adapted technologies. These technologies can be material, such as cyber-physical production systems compatible with a digitalized industrial environment, or immaterial, such as the development of bidirectional digital twins. In this regard, Quentin Charrier (2024) published a comprehensive methodological approach for the design and deployment of a digital twin capable of operating in coupled and bidirectional mode, or in decoupled mode for simulation purposes. In parallel, research conducted by Mohammad Hossein Nikooharf-Tamiz (2024) made it possible to transform a 3D printer into a cyber-physical system through the integration of several sensors and a thermal camera, manipulated by a cobot. This makes it possible to dynamically track the movement of molten filament deposition and collect its temperature in real time.

Algorithmic challenges consist of designing hybrid algorithms that combine conventional numerical methods with artificial intelligence techniques, with the aim of finding optimal solutions to complex computational problems. These algorithms use data collected by cyber-physical systems, enriched with qualitative information such as quality requirements or the various operational ranges. In this regard, Zohreh Shakeri (2025) developed a metaheuristic algorithm dedicated to dynamic operations scheduling. Indeed, the variability of operations and the large number of parameters to consider make resolution particularly complex. Similarly, the research work of Pedram Aminharati (2026) consists of exploiting data from a digitalized 3D printer to develop predictive models capable of adjusting printing speed according to geometric quality requirements. Analysis of the data collected by AI revealed phenomenological correlations between machine vibrations and geometric precision, thus contributing to real-time process optimization.

Finally, the progressive implementation of results from this research work, as well as the acquisition of complementary software and equipment, enabled the Industry 4.0 platform to evolve toward an operating mode considered close to digital. One of the striking indicators of this transformation is that operations supervision is now entrusted to the MES, whereas previously it was the programmable logic controller that controlled the process. Furthermore, the introduction of the bidirectional digital twin is also perceived as a significant advancement for the Industry 4.0 platform.

On the other hand, the use of the term “close” is justified both by the technical constraints encountered and by our choice not to integrate all possible functionalities, as this would have considerably increased the overall level of complexity. This underscores the importance, for an industrial company, of finding a balance between desired functionalities and required investments. Indeed, it has been observed that there is a proportionality between the number of integrated functionalities and the overall level of complexity, which can entail significant hidden costs.

The technical constraints encountered mainly concern limited access to data provided by industrial equipment manufacturers and the ability to activate a manufacturing cycle or any other operation remotely. Indeed, the majority of machines and robots were not initially designed to operate in a bidirectional digitalized environment. Consequently, despite considerable efforts to transform them into cyber-physical production systems, this equipment will never reach the same level of digitalization as those designed from the outset as cyber-physical systems. Nevertheless, current evolution shows that some manufacturers are beginning to relax their restrictions by allowing open access to their machines, under certain conditions. Furthermore, some manufacturers now offer digital twins for their equipment, which constitutes a significant advancement toward more complete digital integration.

From the functionality standpoint, priority was given to three key aspects: flexibility, predictive and portability, considered the most popular in the industrial environment. Flexibility, which is characterized by the platform's ability to manufacture several components of different products, poses numerous challenges such as dynamic scheduling, real-time inventory management, intelligent quality controls and multi-objective optimization. Predictive, for its part, enables the platform to anticipate possible failures or a decrease in measurable performance.

Finally, portability mainly concerns the platform's ability to offer remote and shared interaction with the various authorized actors. The bidirectional digital twin is one of the technologies that facilitates the sharing of information on the progress of operations in real time and offers the possibility of evaluating different scenarios in an immersive manner. The bidirectional aspect makes it possible, under certain security conditions, to receive instructions or execute remotely commanded actions, notably localized diagnostics or an inspection of the production line. Furthermore, the entire industrialization phase preceding production can be simulated thanks to the digital twin and predictive models.

Since 2018, research supported by industrial and institutional partners has made possible the digital transition of the Industry 4.0 platform. This transformation is clearly illustrated by its new operating mode and the various functionalities it offers. Figure 5.2 highlights this evolution, notably through the integration of new cyber-physical production systems such as a mobile robot, a 3D scanner and an IoT system including acoustic, vibratory, optical and energy sensors visible on the lathe (bottom, right). Thus, during manufacturing, data generated in real time is simultaneously recorded in the data lake and displayed near the machine (top, right). Advanced engineering studies were also necessary, thus enabling the robots to autonomously execute loading and unloading operations of parts in the machines.

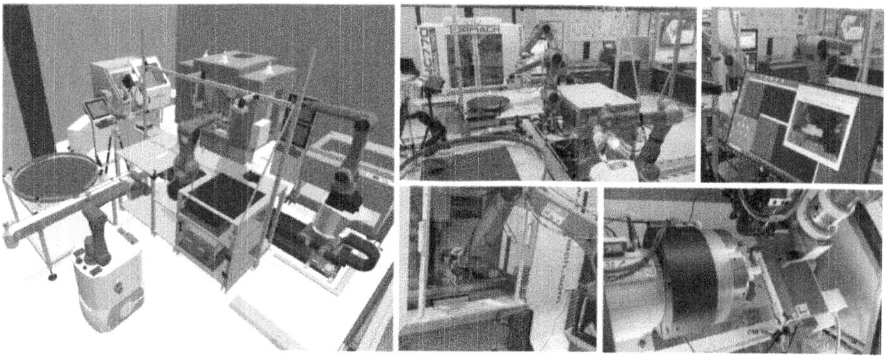

Figure 5.2. *Platform 4.0 – Zone 1. For a color version of this figure, see www.iste.co.uk/benfriha/industrialprocesses.zip*

Furthermore, the remote screens, visible at the top right of Figure 5.2, enable complete integration of the production line. Thanks to these screens, the robot can launch a manufacturing process by pressing the start button on the touch screen using a stylus specially designed for this purpose. Finally, the bidirectional digital twin, illustrated on the left of the image, was successfully tested, paving the way for future technological and technical challenges, particularly regarding its ability to adapt to any evolution, whether material, functional or related to the platform's networks. However, as with any research project, there remains work to be accomplished on new functionalities or innovative technologies to be integrated.

The expansion of the Industry 4.0 platform led to an operational reorganization into three distinct zones. The first zone, the largest in terms of surface area, is dedicated to the production line, as shown in Figure 5.2. The second zone is reserved for assembly, where two cobots collaborate to execute manipulations, as shown in Figure 5.3. The transfer of components between zone 1 and zone 2 is ensured by a mobile robot. Finally, the third zone is dedicated to supervision and is equipped with computers and control screens.

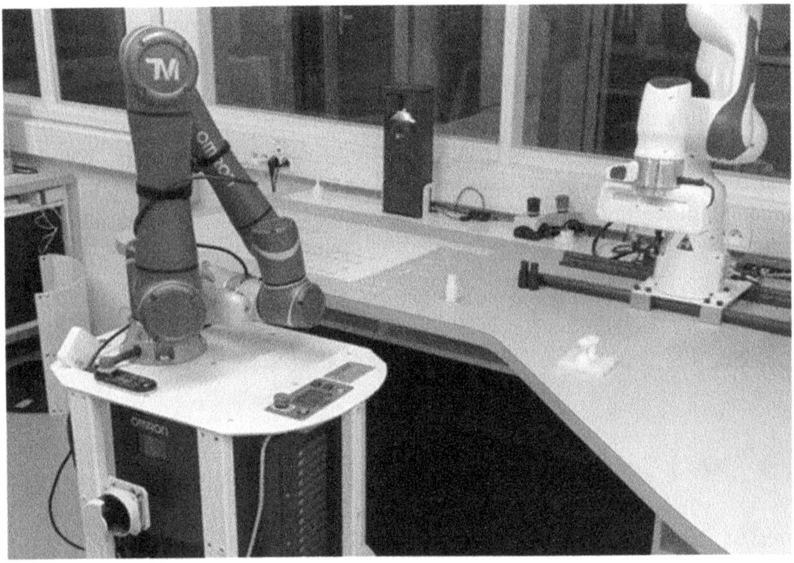

Figure 5.3. *Platform 4.0 – Zone 2. For a color version of this figure, see www.iste.co.uk/benfriha/industrialprocesses.zip*

5.1.2. *New design approach*

The experience accumulated through the various research work and engineering studies conducted on the Industry 4.0 platform led to the creation of a synthesis concept that models a digitalized production line. In Figure 5.4, this concept is structured in four layers that converge toward a supervision cell, both local and remote. Its design is based on the principles of vertical and horizontal integration, while relying on a hybrid pyramid architecture, presented and developed in Chapter 4. Thus, the layers are presented separately to facilitate understanding, but in reality, they are interconnected and interwoven and cannot be thought of independently.

The first layer, physical, is dedicated to machines and their evolution toward cyber-physical production systems (CPS). This is manifested by industrial equipment with single or multiple processes, integrating sensors, actuators, interconnected communication modules, as well as a digital model. Depending on the required level of flexibility, CPS can be adapted to be maneuvered by autonomous mobile robots. This approach makes the production line configuration more agile, thus maximizing operational fluidity.

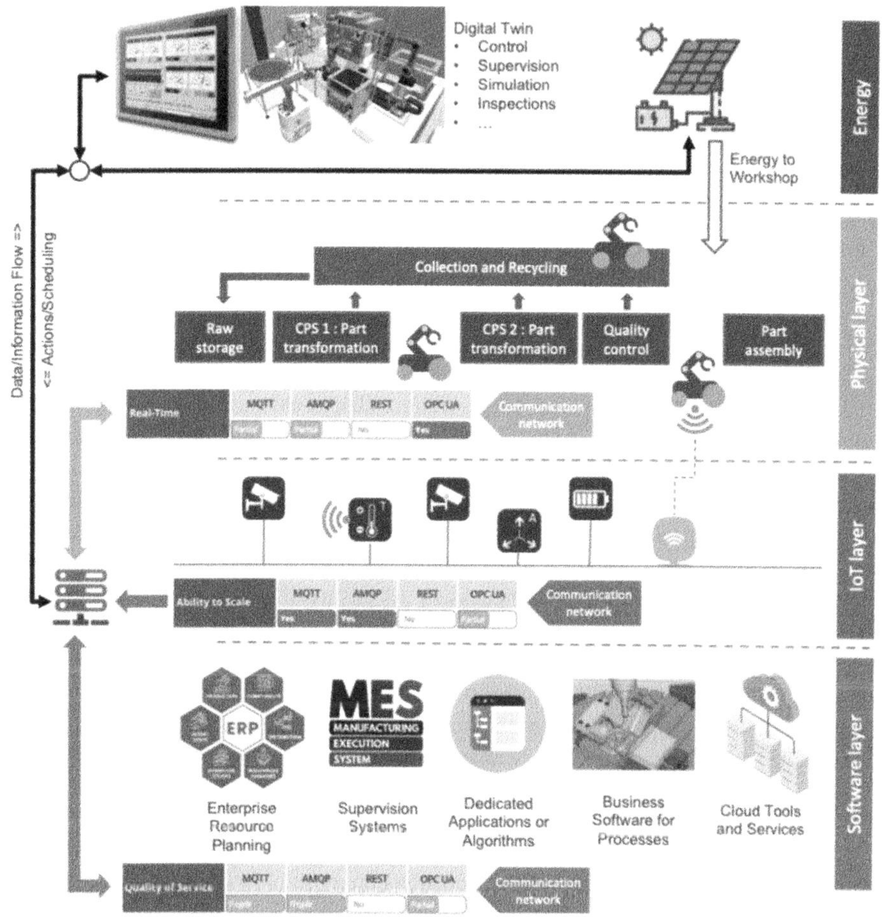

Figure 5.4. *Overall layout of the Industry 4.0 platform. For a color version of this figure, see www.iste.co.uk/benfriha/industrialprocesses.zip*

This layer also includes processes for local recycling, taking into account production waste management. By integrating these practices, companies can reduce their environmental impact while optimizing resource utilization. Local recycling aims to transform residual materials into reusable resources, thus minimizing waste sent to landfills and reducing costs associated with waste disposal. Thanks to Industry 4.0 technologies, for example, algorithms can analyze cutting patterns to minimize waste, while IoT sensors can track and report in real time the areas where material losses are greatest.

The second layer integrates additional sensors of various technologies, the choice of which depends on the targeted functionalities. Known as the "IoT layer", it enables the collection of data and complementary information concerning the progress or status of the process. These sensors use one or more communication networks that are added to the network dedicated to the CPS layer. The nature of the collected data and its storage and preprocessing requirements determine the configuration of the communication networks. In general, the superimposition of communication networks is essential, even if it entails inherent complexity. The diversity of data and their specificities require the use of adapted communication networks and protocols. This makes it possible to fully exploit the collected data, as well as to ensure dynamic real-time supervision.

The third layer focuses on the software hierarchy crucial to the proper functioning of the platform. The integration of commercial software, internally developed applications, cloud services and supervision systems can conceal an intrinsic difficulty. Each software tool has its own requirements, data formats, communication protocols and security standards. It is essential to understand the functions of these tools and their interactions with other components to successfully integrate them. Furthermore, it is necessary to ensure that they can communicate effectively, even if they come from different publishers. Scalability, or the ability to manage an increase in data volume and the addition of new services without compromising performance, must also be verified. Moreover, software tools require regular updates, indispensable for maintaining an acceptable level of security and introducing new functionalities. Managing these updates without disrupting the platform's operation constitutes a real challenge.

On the other hand, publishers often design tailor-made solutions to meet specific industry requirements. However, these solutions are not always designed to integrate easily with other applications, which can restrict their interoperability and integration into varied technological environments. While some publishers favor an openness policy, others opt for proprietary solutions, relying on a closed ecosystem to better control and protect their intellectual property. Publishers who choose openness often seek to stimulate collaborative innovation, attract a wider community of developers and adapt quickly to technological evolutions. In contrast, those who prefer proprietary solutions highlight advantages such as increased security, stability and the possibility of offering exclusive functionalities.

In all cases, the development of internal complementary solutions, so-called "palliative" solutions, often becomes necessary. These are small applications coded in Python, for example, that can serve as gateways to ensure communication between two software programs or with supervision systems.

The last layer addresses the energy aspect. Indeed, energy efficiency has become essential for energy-intensive companies, particularly those operating in the manufacturing, chemical and information technology sectors. Faced with the constant increase in energy costs and the pressure to reduce carbon emissions, these companies must imperatively adopt energy efficiency practices. By optimizing their energy consumption or by recovering waste energy, for example, they can not only considerably reduce their operational costs but also benefit from various tax incentives and government subsidies. Furthermore, industrial digitalization offers new opportunities to monitor, control and optimize energy use.

Thanks to artificial intelligence algorithms integrated into MES, energy-intensive operations can be scheduled to run during periods when the output of photovoltaic panels is maximal. This approach makes it possible to reduce the carbon footprint of production processes and, by extension, of finished products. It thus responds not only to increasingly strict regulations but also to the growing expectations of consumers and investors regarding environmental responsibility. By adopting these strategies, companies can not only improve their ecological performance but also strengthen their competitiveness in the market.

The following chapter will detail the various components, machines and industrial equipment, explaining their evolution toward cyber-physical production systems. This transformation is a prerequisite for the transition toward digitalized production systems. The objective is to propose pragmatic solutions for companies wishing to integrate the digital transition into their development strategy without having to completely replace their industrial infrastructure.

The digital transition must be perceived as a means of producing differently, while satisfying new societal requirements. It is not necessarily about investing in new industrial facilities designed from the outset as cyber-physical systems, which is often unfeasible for most companies due to high costs, associated risks and lack of maturity at different levels, particularly in terms of skills, suppliers and software tools.

It is crucial to note that the digitalization of industrial systems has radically modified the way operations supervision is apprehended. This transformation has had a significant impact on the preparatory phases of manufacturing, notably on programmable logic controller programming, robot trajectories and parametric operational modules. These evolutions are necessary to achieve dynamic operations supervision, capable of adapting scheduling in real time. Finally, all sensors will be integrated into the IoT layer, with detailed information on their implementation.

5.1.3. *Physical layer*

In general, the physical layer is composed of machines, robots, storage areas, as well as various industrial equipment such as a scanner or a 3D printer for local manufacturing of spare parts. The Industry 4.0 platform, for its part, consists of three CNC machines dedicated to manufacturing, including one using a non-conventional process; three robots, including one capable of moving in accessible areas inside and at the periphery of the production line; an assembly area with two cobots; and finally storage areas dedicated to the different stages of the production process. Here is the detail for each piece of equipment:

– a three-axis CNC milling machine (TORMACH PCNC 770) to which two cylinders were added for opening and closing the doors. Furthermore, a pneumatic woodpiece chuck was installed, thus enabling partial automation of the machine. Finally, a complex device was designed and deployed to replace the operator, whose presence was previously necessary to load and launch a machining program or perform other manipulations via the control panel. To this end, the control panel screen was relocated to a touch screen, installed nearby in the robot's workspace. The idea is to equip the robot gripper with a stylus and teach it to mimic the operator, thus reproducing the gestures of the index finger pressing on the screen to load the G-code or launch a machining sequence. This device facilitated the integration of a CNC machine into a digitalized production ecosystem, supervised by the MES. The following two CNC machines benefited from the same device;

– a two-axis CNC lathe (TORMACH 15L Slant-PRO) equipped with a device similar to that mentioned previously for the CNC milling machine. Furthermore, a complete IoT system was designed and installed on this machine to collect and analyze data during the various manufacturing operations. The objective is to develop, through AI learning tools, models of phenomenological behaviors. These models will be used both in the preparation phase to predict behavior according to cutting conditions and thus prioritize parameters to optimize. Similarly, these models will also be exploited during manufacturing in order to alert the MES of any deviation and avoid the production of non-conforming parts. This work is crucial and necessary because in an integrated production system, the monitoring of several production plans in dynamic mode is entrusted to the MES, which needs piloting assistance algorithms capable of analyzing and interpreting upstream data in real time to make objective decisions or order justified actions;

– a CNC laser cutting machine (Universal Laser Systems PLS6.75) equipped with two pneumatic cylinders added to automate the opening of the hood, thus enabling the robot to deposit and retrieve parts after cutting or marking operations. Furthermore, two sensors were added: one to measure the temperature in the laser's workspace and the other to detect the height position of the table (Z-axis). The objective is to secure laser operations through continuous temperature control and to prevent any collision between the part and the laser head. These transformations mark a modest evolution of this industrial equipment toward a digitalized version, thanks to which it will be supervised and monitored via the MES;

– a 3D optical scanner (FARO) used for quality control of machined parts, including geometric and dimensional aspects. Installed near the KUKA mobile robot, in order to place the part on the scanner's rotary table and retrieve it at the end of the operation. The choice of contactless quality control aims to make the production line autonomous and fully integrated. However, it is crucial to take into account certain technical constraints, particularly in terms of lighting, recalibration and materials. Finally, the analysis of 3D scanning data, via tools such as Geomagic, makes it possible to identify dimensional and geometric deviations between the 3D scans and CAD models. Thanks to applications for tracking and recording the observed deviations, the MES can supervise quality by monitoring the evolution of potential drifts and consequently take proactive measures before they leave their tolerance interval;

– a fixed six-axis robot (KUKA KR 900) with an embedded intelligent camera (Sensopart VISOR Allround V20) mounted at the end of the robot, before the gripper. The fixed robot has access to the initial store, the transition store and the conveyor. From the standpoint of elementary characteristics, this robot has a lifting capacity of 6 kg, a repeatability of ± 0.03 mm, a short work cycle and a maximum reach of 900 mm;

– a mobile six-axis robot (KUKA KR 1100) with an embedded intelligent camera (Sensopart VISOR Allround V20) mounted at the end of the robot, before the gripper. This robot is installed on a 3-meter rail, offering it limited but very useful linear mobility. It can thus access three machines, the 3D scanner and the transition store. In terms of elementary characteristics, this robot has a lifting capacity of 10 kg, a repeatability of ± 0.03 mm and a maximum reach of 1,100 mm. However, despite undeniable advances in the design of flexible grippers, these remain below the expectations of production lines where different products of irregular shapes, materials and dimensions are manufactured;

– a mobile manipulator, which unlike the previous mobile robot, moves freely on wheels and offers greater flexibility. The OMRON mobile manipulator is composed of three elements: an LD-90 mobile platform with two LIDARs to determine the various travel trajectories, on which the control device is mounted, and at the top is fixed a six-axis collaborative robot, equipped with an integrated vision system and a flexible two-finger gripper. In this device, video streams are used exclusively for the internal needs of the robot manipulator and cannot be exploited by the platform's supervision system. It is therefore a system that the manufacturer has chosen to lock;

– two Franka Emika Panda cobots, which are installed on rails fixed to an assembly plate (Zone 2). These rails make it possible to vary the distance between the robots, thus offering various usage configurations, including the possibility of bringing the cobots together for collaborative assembly. The assembly plate is located in a room 5 meters from the production line. Once the manufactured parts are available on the accumulation table (Zone 1), their transfer to the assembly plate is ensured by the mobile manipulator. The Panda robot is a seven-axis robotic arm, offering a payload of 3 kg and a reach of 850 mm. Its repeatability is 0.1 mm. Furthermore, the Panda is equipped with integrated sensors, making it sensitive and enabling it to work safely with humans. Learning for the Panda is intuitive, with a touch interface allowing visual coding of tasks;

– an initial store, which is a storage space for raw or finished parts. For the incoming flow, this space serves to store raw parts before their routing to the transition store. This store is generally loaded manually or by the mobile manipulator;

– a transition store, positioned at the center of the Industry 4.0 platform, occupies a crucial place in the execution of production plans. It is a temporary storage space between two operations. Its management relies on complex analytical algorithms, taking into account the presence of different parts at various stages of the production process, as well as the need to manage the geographical and temporal traceability of parts throughout their manufacturing journey;

– a conveyor is a conventional industrial equipment that transports finished parts to a rotary accumulation table. It is very lightly instrumented, providing only binary data concerning its operation, namely start/stop, as well as an optical counter indicating the number of parts detected in transition on the conveyor;

– an accumulation table is positioned at the end of the production line. The conveyor discharges the manufactured parts, conforming to the quality requirements associated with them, onto the rotary accumulation table. The extraction of parts from this table can be done manually or via a mobile manipulator which, thanks to its camera, locates and retrieves the parts to route them to the assembly site.

5.1.4. *Software and control layer*

The software and control layer consists of various software programs and applications. Some are essential, as they integrate intelligent algorithms, while others play a complementary role by acting as gateways between different communication entities. Only the most significant software programs will be presented below:

– the MES, which is composed of several service-oriented applications, notably the Aveva System Platform integrated development environment (IDE), the integration server, the HMI developed by Intouch OMI and the historian. A Domain Name System (DNS) server is installed in order to activate the Intouch Access Anywhere application, playing the role of a remote HMI application for portable use;

– the TIA Portal, which makes it possible to reprogram the logic functions of the Siemens Programmable Logic Controller (PLC) S7 1500 that acts as a moderator and distributes in a standardized manner the variables (inputs and outputs) between the MES and the machines. The introduction of the MES into digitalized production systems imposed a new command flow, where manufacturing orders and their scheduling are managed and transmitted by the MES to the machines. Thus, the role of the PLC is now limited to serving as a gateway between the MES and the machines;

– the digital twin, which is a digital copy of the platform, was developed using Unity3D and integrated into the system. The 3D modeling and kinematics of the robots were developed through KUKASim and then exported to the Unity3D project. This bidirectional digital twin makes it possible to supervise an entire manufacturing process. Similarly, it can display information on the progress of operations or launch an inspection plan, for example;

– production preparation tools, which in the context of industrial process digitalization are essential. Dynamic operations scheduling requires the decomposition of the production plan for each part into several parametric modules. To this end, specialized business software is used to create modules comprising parametric G-codes and additional lines of code, enabling dialogue with the environment, such as checking the status of machine doors. The MES is responsible for organizing the modules over time and assigning values to parameters, notably execution speed, according to optimization priorities defined by production managers;

– cloud computing tools, which offer a range of varied services, from scalable data storage capacity to powerful computational resources for advanced analyses. They also enable secure data sharing. Furthermore, the cloud supports modeling, simulation and integration of IoT technologies, while offering an interface and tools for designing specific algorithms. All of these services contribute to the overall optimization of production processes.

5.1.5. *IoT layer*

This section presents feedback on the deployment of the IoT layer, including sensors and cameras integrated into the Industry 4.0 platform. The system was tested under real manufacturing conditions, and its main components are presented below:

– An inspection camera installed overhead, which offers an overall view of the platform. It is used for surveillance purposes, particularly to ensure the safety of persons present. Indeed, several researchers, doctoral students and interns collaborate on the implementation and experimentation of Industry 4.0 technology building blocks. Thanks to this camera and image analysis, the MES can verify whether a person is near the machines and robots before launching any operation.

– A fixed camera system, installed above the initial storage area and the intermediate storage area. Their main mission consists of monitoring in real time the movements of parts in these storage areas and transmitting this information to the MES. This includes referencing, traceability, part status and the dynamic temporality of movements.

– A computer, installed near the production line, which is dedicated to computing, storage and data collection. It is particularly used for surface reconstruction from point clouds obtained by part scanning. This computer has a 20-terabyte hard drive, a 14-core 3.30 GHz processor, 2 graphics cards with 22 GB of VRAM and 128 GB of RAM. The latter is scalable and allows the integration of new components.

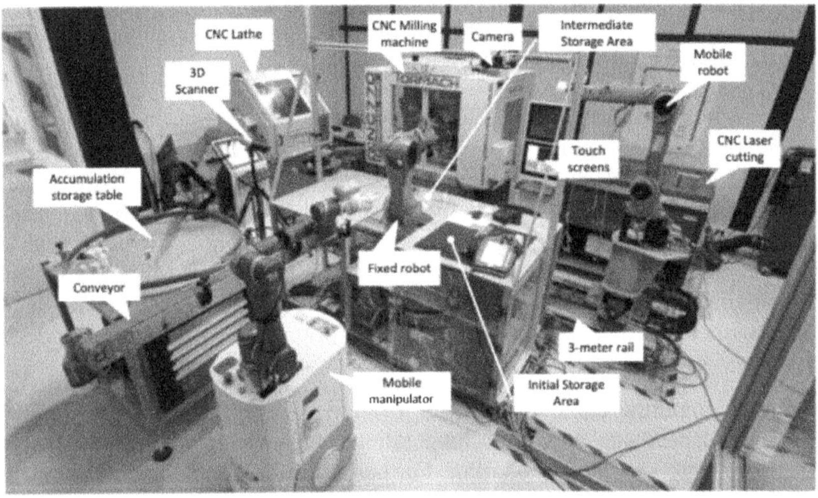

Figure 5.5. *ENSAM's Industry 4.0 platform. For a color version of this figure, see www.iste.co.uk/benfriha/industrialprocesses.zip*

Regarding sensors, all machines and robots are equipped with energy sensors making it possible to measure the energy consumption of each piece of equipment. Furthermore, an IoT box, comprising temperature and table position sensors on the Z-axis, is installed on the laser cutting machine. The lathe, for its part, is equipped with several sensors due to numerous research focused on the development of predictive models aimed at anticipating a degradation of product quality or of the manufacturing tool. Here are the sensors that were implemented:

– A camera installed inside the lathe, which captures photos and videos of the part during machining. An initial visual analysis of these images revealed that the initial position of the part is inclined relative to the Z-axis of rotation of the chuck. This information is crucial for the manufacturing process. This inclination is probably caused by several technical factors, such as the offset between the part reference frame and that of the robot gripper. Another factor could be a lack of robot precision when inserting the cylindrical part into the chuck, where the cylinder axis does not align with the chuck's Z-axis. In any case, this technology made it possible to detect a malfunction previously ignored, paving the way for new studies to find solutions. The latter could not be observed by operators due to the "integrated" aspect of the production line. Indeed, as mentioned previously, the MES is programmed to inspect the production line before any start-up, and in the event of human presence in the manufacturing cell, launch is suspended.

– Two vibratory sensors (an accelerometer for force measurement and a gyroscope for rotation measurement), which are installed on the block supporting the chuck's rotating shaft, near the bearings, to measure the vibrations generated. The capture and analysis of vibratory signals is not new. This technology is widely used in preventive and condition-based maintenance of rotating machinery. In this experiment, these vibratory measurements, recorded during manufacturing, are analyzed by an AI algorithm in a context extended to the entire production line and its environment. The objective is to identify and characterize phenomenological correlations between different physical elements, impacting part quality or tool condition. Machine degradation or aging was excluded from this approach due to the recent state of the machine fleet, as well as to limit the complexity of computational models.

– A microphone, which is installed inside the machine to measure acoustic signals near the part and the tool. In the workspace where machining takes place, various movements are carried out, notably rapid or slow movements to bring the tool closer to or away from the part, spindle rotation at different speeds, the impact between the tool and the part, as well as the material removal sequence with chip generation. Each movement or contact produces a specific or combined acoustic signature, the analysis of which can reveal valuable information on cutting conditions, tool wear or any other operating mode associated with a specific acoustic signature. However, a manufacturing workshop is generally surrounded by machines, robots and industrial equipment generating noises likely to disturb localized listening. For this reason, active sound filtering algorithms can be deployed to make the analysis of captured acoustic signals relevant.

After having presented all the machines, robots, IoT devices, control systems, storage areas and sensors integrated into the Industry 4.0 platform, section 5.2 will address the data flow between these various machines, detailing the communication protocols used.

5.2. Communication architecture

Communication architecture in a connected factory is fundamental in the operation of advanced functionalities and data valorization, whether through on-site systems or via cloud computing. Furthermore, the availability of real-time data makes it possible to orchestrate operations dynamically, thus conferring upon production systems increased agility and responsiveness.

This section presents the current IT architecture of the Industry 4.0 platform, which is composed of two parts. The first part, shown in Figure 5.6, offers a detailed overview, organized into three distinct zones. This division into zones is crucial for several reasons: it simplifies the overall understanding of the architecture by offering a functional perspective, facilitates the integration of new cyber-physical systems or software, and makes it possible to anticipate the cybersecurity needs of the site. This involves measures such as zoning, which groups critical components, such as programmable logic controllers and switches, in a secure zone, where their protection is reinforced by devices such as access badges and intelligent cameras.

The second part focuses on the eight communication protocols used between the various machines, robots, sensors, programmable logic controllers, cloud, as well as the supervision and control system. As mentioned previously, the different transition phases of the Industry 4.0 platform have complexified the IT network, particularly due to the diversity of communication protocols and their coexistence, which requires the use of specific gateways. This makes it possible, on the one hand, to guarantee data fluidity and integrity, and, on the other hand, to ensure the transmission and execution of actions ordered by the MES.

5.2.1. *Connection architecture*

Since the launch of the Arts et Métiers "Industry 4.0 platform" project, the architecture has evolved through several stages to adapt to the requirements of the digital transition, thus integrating technological advances in line with the targeted functionalities. Initially, the architecture was centered around a PLC (Programmable Logic Controller), designed to execute pre-established production plans, without the slightest operational variability. This PLC was equipped with a human–machine interface (HMI), making it possible to visualize the state of logical variables and assign values to production parameters, such as the number of parts to be produced. This operating mode offered little flexibility. Today, the transition process has resulted in a complete overhaul of the Industry 4.0 platform architecture. Although it has become complex, it now offers new functionalities, such as flexibility, prediction and autonomy, globally improving performance.

The new connection architecture, shown in Figure 5.6, is based on a virtual local area network (VLAN), which serves as a junction point between the three zones. This configuration enables an active connection between Zone 3 and the two production zones, namely Zones 1 and 2.

Zone 1 is designed to offer both dynamic operations supervision and control, as well as tools adapted to the industrialization phase within the framework of digitalized production. This involves decomposing the production plan into parametric modules, which makes it possible to render operations orchestration flexible, regardless of the variability of products being manufactured. Furthermore, thanks to the digital twin, validation of the production plan can be simulated in an immersive and interactive manner. This approach represents a break from traditional systems.

Zone 2 underwent a profound transformation. Existing industrial equipment evolved toward cyber-physical production systems to be compatible with a digitalized ecosystem. This transition required several complementary works.

For example, the automation of doors, the addition of connected sensors and the installation of three touch screens. These screens make it possible to relocate certain actions, such as loading a G-code or pressing the start button to launch a manufacturing cycle.

Previously, these actions required the physical presence of an operator in front of the control panel. Today, these actions are performed by a robot which, using a stylus specially designed for this purpose, reproduces human gestures by pressing on the touch screen, for example, at the location of the start button.

These evolutions had consequences on the wired communication network. The addition of industrial equipment, such as the mobile manipulator or the 3D optical scanner, also required specific connections and protocols. Furthermore, for the energy monitoring of the Industry 4.0 platform, all equipment is connected to the Socomec box, which measures energy consumption variations over time and uses a specific network dedicated to asynchronous transmission.

Zone 3, equipped with recent equipment, benefits from a more harmonious communication network, based on a single protocol: TCP/IP. This configuration shows that industrial equipment suppliers are becoming increasingly aware of the needs of Industry 4.0. They are beginning to integrate these technologies from the design phases, which considerably simplifies their integration into a digitalized production environment.

However, there is still a long way to go for suppliers to be able to offer cyber-physical equipment truly ready to be integrated into a production line in plug-and-play mode. Several technical, normative and strategic obstacles continue to hinder the widespread adoption of these innovative systems. These challenges require particular attention to be overcome and to enable effective integration.

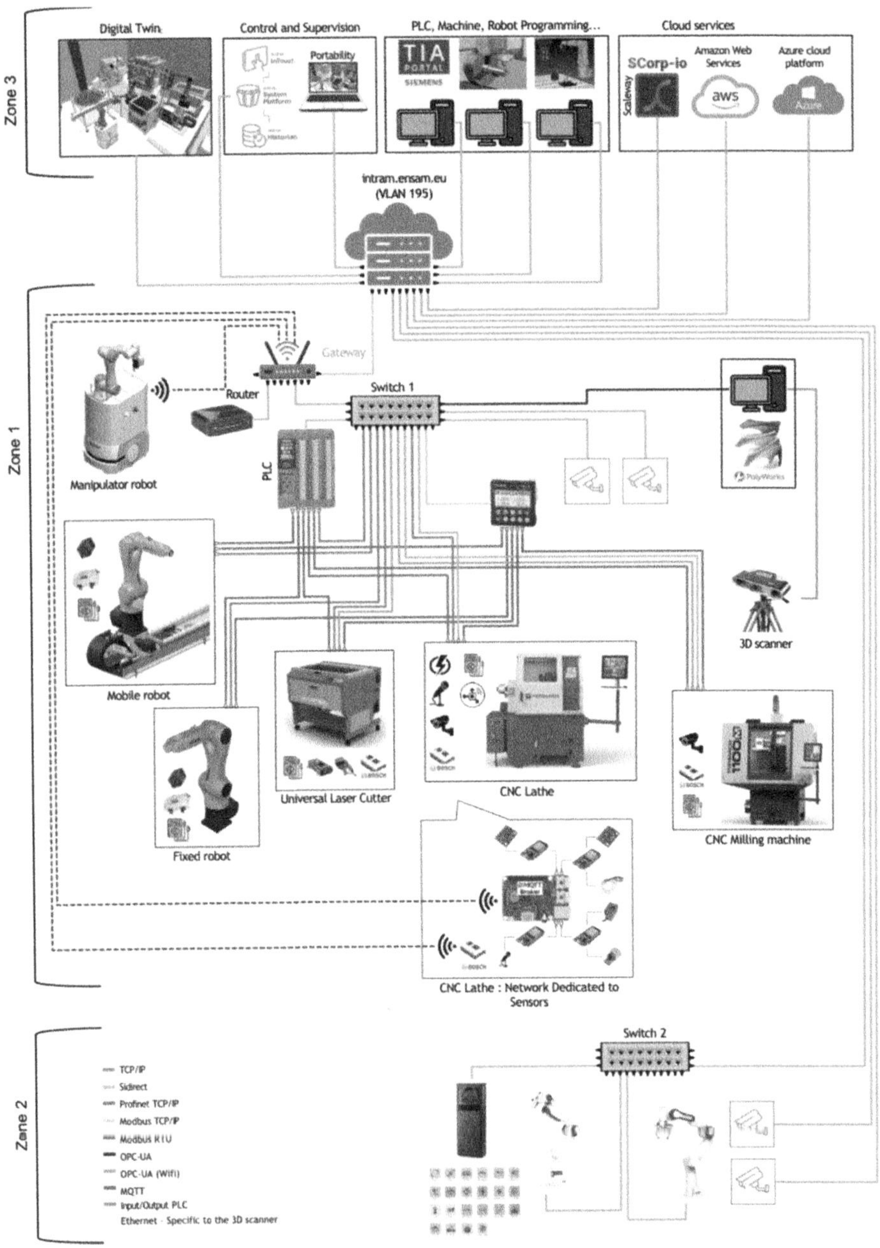

Figure 5.6. *IT architecture of the Industry 4.0 platform. For a color version of this figure, see www.iste.co.uk/benfriha/industrialprocesses.zip*

Thus, machine safety standards often impose the use of communication networks based on the PLCs. These networks are essential for ensuring the safety and reliability of industrial operations, but they can also pose challenges in terms of interoperability and flexibility. For example, traditional protocols such as Modbus, Profibus and Ethernet/IP, although essential for machine-to-machine (M2M) communication, can limit the ability of systems to adapt to the new requirements of Industry 4.0.

Standards such as ISO 13849-1 and the Machinery Directive 2006/42/EC impose strict requirements on equipment safety, which complicates the direct integration of cyber-physical systems. These standards require specific performance levels and in-depth risk analyses, thus increasing the complexity of designing and implementing plug-and-play solutions. Furthermore, the specific needs of end users are difficult for suppliers to anticipate, which makes creating universal and modular solutions difficult, forcing companies to make post-installation adjustments, which increases costs, delays and implementation complexity.

Beyond normative and technical challenges, there are also major strategic obstacles that do not facilitate the widespread adoption of advanced technologies. One of the main concerns of suppliers lies in their reluctance to open access to their machines' data. Data generated by industrial systems is extremely valuable, enabling advanced analyses thanks to artificial intelligence technologies. However, many suppliers fear that the release of this data would primarily benefit digital platform companies, which could exploit it to develop value-added services to the detriment of equipment suppliers themselves. This concern is amplified by rapid advances in the field of data science, where machine learning and deep learning algorithms can transform raw data into actionable strategic information.

To overcome these challenges, it is imperative to promote increased collaboration between equipment manufacturers and digital players. A collaborative approach could include the development of open standards and robust security protocols that enable interoperability without compromising data security. Organizations such as the Industrie 4.0 Working Group and the Open Platform Communications (OPC) Foundation are already working on the development of standards such as the OPC Unified Architecture (OPC UA), which aims to harmonize communication between machines while ensuring data security and confidentiality.

Furthermore, innovation in the field of modular and adaptable solutions offers promising perspectives and could help meet the varied needs of end users. The creation of systems capable of self-configuring or dynamically adapting to changes in production environments represents significant progress.

Finally, it is essential to foster a co-innovation approach among all stakeholders of the industrial ecosystem. This includes not only suppliers and system integrators but also end users and standardization bodies. This collaborative approach could help accelerate the transition toward a truly digital and integrated industry, in accordance with the principles of Industry 4.0.

5.2.2. *Communication protocols*

The TCP/IP protocol (Transmission Control Protocol/Internet Protocol), widely used in the Industry 4.0 platform, ensures communication and interoperability between various equipment. By combining IP for data addressing and routing, and TCP for managing reliable connections, TCP/IP facilitates the integration of technologies such as IoT, the digital twin and process supervision systems. Its main characteristics include reliable communication, interoperability between heterogeneous equipment, as well as efficient congestion and error management. It serves as the basis for many modern industrial protocols, supporting the transition toward digital production environments. However, it does not provide specifications for industrial real-time applications or data models.

SiDirect is a communication solution developed by Siemens, specifically designed for the integration of industrial automation systems. Using standard protocols such as TCP/IP, SiDirect facilitates communication between Siemens equipment and other industrial systems. Its main characteristic is its ability to simplify real-time data exchange and ensure smooth interoperability between various devices, thus enabling centralized process management, better operations synchronization and effective integration into complex production environments.

PROFINET TCP/IP is an industrial communication protocol developed to integrate automation systems via Ethernet networks. Based on TCP/IP, PROFINET offers real-time communication between automation devices, such as programmable logic controllers and sensors of the Industry 4.0 platform, notably for advanced diagnostic functionalities. It is distinguished by its ability to manage data with minimal latency and to support complex distributed networks. PROFINET TCP/IP facilitates interoperability between equipment from different manufacturers and enables effective integration into digitalized production environments.

Modbus TCP/IP is a communication protocol used for data exchange in industrial systems via Ethernet networks. Based on the TCP/IP protocol, Modbus TCP/IP enables communication between devices such as programmable logic controllers and sensors, with a simple and efficient data structure using master-slave mode data transactions and fixed data frames. This protocol is appreciated for its simplicity, its compatibility with various equipment and its ability to provide fast and reliable data exchanges. Although it is suitable for simple applications with less complex communication needs, it is less performant for applications requiring high real-time performance, compared to PROFINET.

Modbus RTU is a serial communication protocol used for data exchange in industrial systems. It operates on serial networks such as RS-232 or RS-485 and is based on a compact and efficient data structure. Modbus RTU is widely recognized for its simplicity, robustness and ability to enable reliable communications between sensors, particularly those measuring energy consumption variability. However, it is more suitable for older or less complex systems, with limited capabilities in terms of speed and complexity compared to Ethernet-based protocols.

The OPC UA is a standardized communication protocol designed for data integration and exchange in industrial systems. It provides a unified architecture that supports secure and interoperable data exchanges between various devices and software. OPC UA is based on modern technologies such as TCP/IP and offers advanced functionalities such as data modeling, integrated security and connection management. It is highly suitable for digitalized environments with high requirements in terms of security, interoperability and data management.

Message Queuing Telemetry Transport (MQTT) is a lightweight communication protocol used for data exchange in M2M communication systems and IoT. Designed for low-bandwidth and high-latency environments, MQTT operates on TCP/IP networks and uses a publish/subscribe model to transmit messages. Its main strength lies in its ability to provide efficient and reliable communication with minimal overhead, making it ideal for industrial applications requiring real-time communication and low bandwidth consumption. MQTT facilitates the integration and management of IoT devices by offering a flexible and scalable solution for connected environments.

The input/output interfaces of PLCs refer to the connections of a programmable logic controller used in industrial automation. Inputs allow the PLC to receive signals from various sensors and field devices, such as switches, temperature sensors and level detectors. Outputs allow the PLC to control actuators and devices such as motors, valves and relays. By managing these signals, the PLC executes control programs to automate industrial processes. Although the PLC is essential for industrial automation, it lacks flexibility.

However, things are evolving with the IEC 61499 standard, introduced in 2005, which redefines the use of PLCs by proposing a distributed, flexible and modular model. Unlike conventional centralized PLCs, this standard makes it possible to distribute controls over several nodes, thus improving scalability and reducing latency thanks to an event-driven approach. Although its potential is recognized, its adoption remains limited among industrial companies, with approximately 10–20% of companies actively using it, mainly in pioneering sectors or for projects requiring great flexibility and interoperability. This slow adoption results from the prevalence of conventional PLCs and the resistance to change observed in some industries, despite the considerable advantages in terms of scalability and responsiveness offered by Industry 4.0. In this context, the research of Bencherki (2023) aims to facilitate the integration of these technologies by developing structuring approaches that are interoperable with other systems already integrated into digitalized production systems, such as MES and the digital twin.

The eight protocols mentioned previously – TCP/IP, PROFINET TCP/IP, Modbus TCP/IP, Modbus RTU, OPC UA, MQTT and the input/output interfaces of PLCs – each possess distinct characteristics that make them more or less suitable for different situations, particularly within the framework of the Industry 4.0 platform. This clearly illustrates the inherent complexity of designing and deploying various networks to achieve the desired objectives.

In summary, each protocol presents its own advantages and limitations. The choice of protocol depends both on the technological state of the existing equipment and on specific requirements such as the functionalities to be integrated, the technological solutions retained (such as the deployment of a digital twin or the integration of a mobile manipulator), the level of complexity, the operating modes (whether coupled or decoupled), the operational dynamics (whether real time or deferred), as well as data management and analysis.

5.3. Generated data

Industry 4.0 relies on the massive integration of digital technologies within production processes, placing data at the heart of this revolution. Smart factories generate, collect and analyze in real time quantities of data coming from machines and IoT devices. The exploitation and valorization of this data within the framework of the Industry 4.0 platform constitute essential levers, underscoring the importance of a rigorous classification approach.

5.3.1. *Data classification*

Data is classified according to two main criteria: category and provenance. Four categories have thus been defined: machine-intrinsic data, sensor-generated data, images and operational data. The collection method varies depending on the source of each data type.

This classification model makes it possible to integrate all data generated on the platform. Furthermore, it is designed to be compatible with cloud computing infrastructures, thus facilitating data migration and analysis thanks to available computational resources. This model also enables local exploitation of collected data. Within the framework of the Industry 4.0 platform, only data produced by three industrial equipment items will be illustrated: the lathe, the laser cutting machine and the 3D scanner.

5.3.2. *Examples*

The first example cited (see Table 5.1) offers a synthesis of data associated with a CNC lathe machine. This machine is equipped with numerous sensors, thus generating a considerable volume of data during manufacturing cycles. Among these sensors are those measuring mechanical vibrations, acoustic signals and images capturing part contours. Some of these data streams require local preprocessing, such as the conversion of signals from the time domain to the frequency domain.

Categories	**CNC lathe machine**	
Machine data	Type	Alarm, communication status, availability and machine status, machine data, I/O modules
	Format	CSV (Comma-Separated Values)
Sensors and IoT	Sensor	Energy sensor, vibration sensor (chuck), vibration sensor (motor), microphone
	Data	Energy consumption, vibration frequency (chuck), vibration frequency (motor), acoustic frequency inside the machine
	Format	CSV (Comma-Separated Values)
Images	Sensor	Camera
	Type	Part contour
	Format	Image
Operation-related data	Type	G-code, cutting conditions, cycle time
	Format	Text

Table 5.1. *Classification of data generated by the CNC lathe machine*

The main objective of local processing is to avoid overloading the platform's communication network with several gigabytes of data. It also makes it possible to quickly transform raw data collected near the machine into actionable information that the MES can integrate into a decision-making process. The information generally highlighted concerns operational anomalies, such as unexpected impacts or unusual sounds. Furthermore, the MES can display this information in the digital twin, making it accessible and exploitable by different actors, whether on-site or remote. Moreover, raw data is stored in its initial format in the data lake for subsequent in-depth analysis, if necessary.

This approach is also applied to other industrial equipment on the platform, such as the laser cutting machine. However, this process does not generate a large quantity of data. Indeed, the operating system is entirely locked by the manufacturer and returns no internal data. In this context, two sensors were installed to improve the active safety of the machine. Thus, the only data transmitted to the MES, at low frequency, concerns the temperature in the chamber where the intense beam is generated, as well as a measurement of the distance between the table and the nozzle. This makes it possible to monitor and prevent any material degradation.

Categories	**CNC laser cutting machine**	
Machine data	Type	Alarm, communication status, availability and machine status, machine data, I/O modules
	Format	CSV (Comma-Separated Values)
Sensors and IoT	Sensor	Thermometer and laser
	Data	Temperature and nozzle height positioning
	Format	CSV (Comma-Separated Values)
Images	Sensor	–
	Type	–
	Format	–
Operation-related data	Type	G-code, laser power, cycle time
	Format	DXF, PDF and EMF

Table 5.2. *Classification of data generated by the laser cutter*

Furthermore, the 3D scanner aims to verify the geometric and dimensional conformity of produced parts, by comparing them to their CAD model. This equipment benefited from a framework developed specifically to establish a bidirectional dialogue gateway with the MES because it was not designed as a cyber-physical system and is therefore not natively compatible with digitalized systems. Quality 4.0 is a complex domain and is the subject of numerous research projects, aiming in particular to develop dynamic, autonomous and intelligent control capable of modeling the variability of dimensional and geometric behaviors, as well as predicting any deviation from the quality reference frames of manufactured parts.

Categories	3D optical scanner	
Machine data	Type	Conforming/non-conforming products, deviation values
	Format	CSV (pass/fail result) and PDF (report)
Sensors and IoT	Sensor	–
	Data	–
	Format	–
Images	Sensor	–
	Type	–
	Format	–
Operation-related data	Type	Scan program, 3D model
	Format	STEP and Python

Table 5.3. *Classification of data generated by the 3D scanner*

5.4. Case study

This case study aims to illustrate how industrial equipment, such as a 3D scanner, can be integrated into an Industry 4.0 platform to create a digitalized device dedicated to quality control. Within the framework of the overall layout of the Industry 4.0 platform (see Figure 5.4), this project is positioned in the physical layer and the software layer.

The main challenge consists of transforming this equipment into a cyber-physical system compatible with the platform's digitalized ecosystem. Indeed, numerous industrial equipment items were initially designed without taking into account compatibility requirements with an advanced digital environment, which often requires engineering work that can prove costly without guaranteeing a complete transformation. Furthermore, the software used for surface reconstruction requires operator intervention and does not return dimensional and geometric conformity results to the MES. This limitation poses an additional challenge in the automation of quality control processes because autonomous quality assessment and monitoring are essential functionalities expected from digitalized industrial systems.

Before detailing the transformation of a 3D scanner into digitalized equipment capable of interacting with the MES, an introduction to the concept of Quality 4.0 as well as an analysis of the technological choice used to scan components will be presented.

5.4.1. *Introduction to Quality 4.0*

With the advent of Industry 4.0, the concept of Quality 4.0 has emerged to redefine traditional approaches to quality management within digitalized production systems. This new approach integrates technological advances that make it possible to manage quality in line with the requirements of new functionalities, such as speed, autonomy and flexibility. The main objective of Quality 4.0 is to dynamically control product conformity at each stage of their manufacturing, while minimizing the impact on production cadence.

Quality 4.0 involves the digitalization of quality control processes, which transforms the way companies approach quality management. This transformation relies on the integration of digital technologies and compatible equipment. Thus, 3D scanners represent a contactless measurement solution, which makes it possible to render the control process more agile. For example, the robot can move the scanner inside the machine to perform in situ measurements, without having to extract the part. This makes it possible, on the one hand, to save operational time, and, on the other hand, if a non-conformity is detected, the MES can choose to suspend the manufacturing process of the concerned part or adjust the G-code to compensate for the measured error.

Furthermore, Quality 4.0 is not limited to conformity measurement aspects; it relies on other technologies to optimize various aspects of manufacturing processes. IoT sensors, for example, collect real-time data on machine and product status to prevent defects before they affect final quality. Similarly, artificial intelligence enables in-depth analysis of a vast quantity of data in order to establish potential correlations between components deemed non-conforming and events that occurred during the manufacturing process. These capabilities make it possible to suggest actions to maintain constant quality.

Likewise, blockchain ensures transparency and traceability in data transactions, thus guaranteeing that all production stages meet defined quality standards. In addition, augmented reality (AR) and virtual reality (VR) offer immersive environments to train operators in 3D scanner adjustments, notably for calibration, positioning and orientation of optics. Although this technology is advantageous due to its contactless nature, it can be affected by ambient light. Indeed, certain materials or surface states can react to the light projected by the scanner, leading to the formation of a point cloud that does not accurately reflect the scanned geometry.

Furthermore, Quality 4.0 is not limited to conformity measurement aspects; it relies on other technologies to optimize various aspects of manufacturing processes. IoT sensors, for example, collect real-time data on machine and product status to prevent defects before they affect final quality. Similarly, artificial intelligence enables in-depth analysis of a vast quantity of data in order to establish potential correlations between components deemed non-conforming and events that occurred during the manufacturing process. These capabilities make it possible to suggest actions to maintain constant quality.

Likewise, blockchain ensures transparency and traceability in data transactions, thus guaranteeing that all production stages meet defined quality standards. In addition, augmented reality (AR) and virtual reality (VR) offer immersive environments to train operators in 3D scanner adjustments, notably for calibration, positioning and orientation of optics. Although this technology is advantageous due to its contactless nature, it can be affected by ambient light. Indeed, certain materials or surface states can react to the light projected by the scanner, leading to the formation of a point cloud that does not accurately reflect the scanned geometry.

Quality 4.0 makes it possible to monitor, improve and predict product quality in real time throughout the production chain. It offers integrated visibility and control, which enables companies to detect and resolve quality problems before they affect final products. This approach promotes proactive management, thus reducing risks related to quality defects.

The implementation of Quality 4.0 in a digitalized production system presents numerous advantages. It improves overall performance through automated systems and advanced analyses, enabling early detection of deviations and minimizing post-production corrections. Furthermore, integrated monitoring reduces downtime, promoting continuous production. This flexibility also makes it possible to meet traceability and precision requirements, while reinforcing customer confidence through constant quality. In summary, Quality 4.0 is essential for maintaining efficient production flows and optimized performance.

5.4.2. *Technology used*

The choice of a 3D scanner is justified by its ability to accurately measure physical objects, thanks to contactless digitization. A 3D scanner can also be manipulated by a robot, which moves it and sweeps the object to be scanned, thus offering superior flexibility compared to traditional contact measurement methods. These characteristics make it a tool compatible with digitalized industrial processes.

Scanners collect and store detailed data on the size and shape of the object. In particular, structured light scanners operate by capturing a series of images while different light patterns are projected onto the surface of the stationary object, which makes it possible to create a complete 3D point cloud of its geometry. Then, the geometric and dimensional conformity verification process relies on sophisticated computational algorithms, which compare, for each scanned part, the deviations between the 3D model reconstructed from the point cloud and the native model from CAD. This ensures that each component meets the required specifications.

When choosing a 3D scanner, several essential factors must be taken into account. This includes data format compatibility, accuracy, resolution, scanning speed and portability, particularly if the scanner is intended to be manipulated by robots to perform in situ verifications.

Furthermore, the cost of the scanner is a factor to consider. It is crucial to compare not only the initial purchase price but also the overall costs, including acquisition, integration, software updates and technical support.

In the case of the Industry 4.0 platform, the choice was made for a structured light scanner that perfectly meets the needs of the Industry 4.0 platform: the FARO Cobalt Array Imager. This scanner is designed for use in various fields, such as quality control, inspection and reverse engineering. It offers a resolution of up to 0.1 mm and can capture up to 1.5 million points per second.

Furthermore, thanks to its wide field of view and portability, the FARO Cobalt Array Imager can be used in varied contexts.

5.4.3. *General approach*

The objective of this study is to integrate the 3D scanner as a cyber-physical module, dedicated to quality control and supervised by the MES. This module will have the function of collecting data, analyzing measurements and their trends, and suggesting proactive or corrective actions in the event of deviations.

To achieve these objectives, a methodology structured in five steps is implemented, as shown in Figure 5.7.

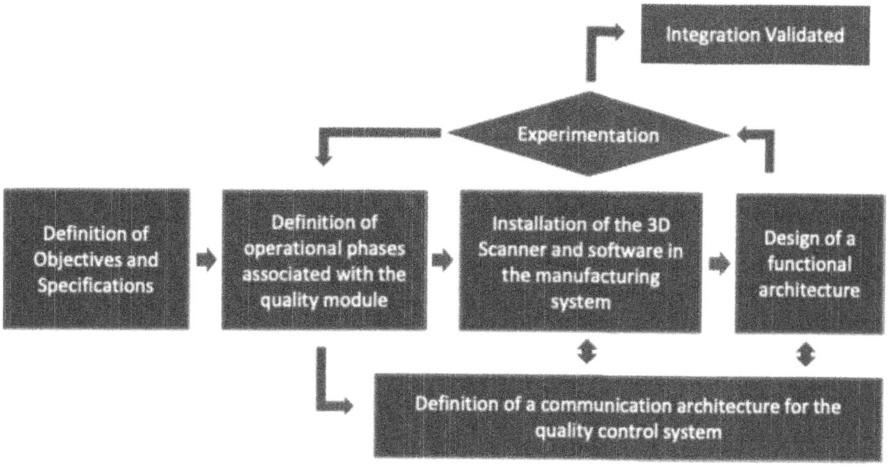

Figure 5.7. *Methodology for integrating a quality control CPS module*

5.4.3.1. *Step 1 – definition of objectives*

The first step consists of defining the integration objectives of a module dedicated to Quality 4.0 and determining its optimal location in the Industry 4.0 platform, while taking into account its accessibility by robots. The objective is to make quality control compatible with a digitalized manufacturing process, emphasizing fluidity, autonomy and traceability of operations. Furthermore, advanced analysis of data collected by 3D scanners can provide information, not only on the conformity of parts with quality reference frames but also on potential correlations between the quality variation of certain parts and the simultaneous occurrence of events on the Industry 4.0 platform. For example, analysis by artificial intelligence can reveal temporal similarities between the quality degradation of certain parts and the startup of a distant machine generating significant vibrations.

5.4.3.2. *Step 2 – definition of operational phases*

Next, the second step focuses on identifying the operational phases introduced by this new module, as well as the types of incoming and outgoing data or information for each of the operations. This will make it possible to determine "where" and "when" communication is necessary and to define the appropriate communication protocol, as well as the essential variables for each operation. In this perspective, the first version proposes to limit the number of phases to three, as shown in Figure 5.8: the preparatory phase, the active phase and the decision phase.

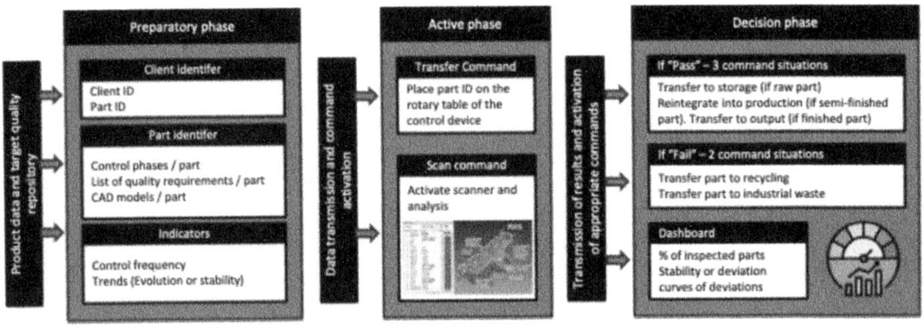

Figure 5.8. *Data required in the different phases of the quality control process*

The preparatory phase consists of verifying the availability of all information necessary for the progress of the quality process. This includes identifiers used by the MES, native CAD models, as well as the quality repository specifying the dimensional and geometric characteristics to be respected, with their tolerances. Furthermore, it determines the indicators to be displayed on the dashboard, such as inspection results, represented here by a binary pass or fail variable, and the rate of inspected parts, which can reach 100% if the customer wishes to verify all parts. Finally, continuous analysis of measurements can reveal degradation or stabilization trends, expressing the evolution of deviations between measured values and reference values, thus making it possible to anticipate potential non-conformity situations.

Concerning the second phase, which focuses on the quality control process, it consists of two main commands. The first, called "transfer command", aims to ensure the robotized movement of parts to be verified to the scanner's rotary table. The second command launches the point cloud capture process thanks to the simultaneous rotation of the table and image capture, followed by surface reconstruction and a comparison with the CAD model. Finally, a global conformity report is generated, indicating pass or fail according to the collected measurements and their tolerances.

The third operational phase focuses on the decisions that the MES can take according to the obtained results. In the event of conformity, the MES reintegrates the part into the production process by activating the dynamic scheduling algorithm for this purpose. In the event of non-conformity, the part is redirected to recycling or industrial waste. After each inspection, the indicator dashboard updates systematically, providing a synthetic view of the monitoring parameters for the quality of produced parts.

5.4.3.3. *Step 3 – physical installation of the scanner*

The third step consists of installing the necessary hardware and software. The scanner or physical module must be fixed in an appropriate location on the production site and connected to the MES. It is important to consider the accessibility of the device, particularly for robots or mobile manipulators, in order to guarantee operational autonomy and smooth integration into the planning of production operations.

Other technical constraints may appear, such as the lighting of the scanning area or the orientation of the component, in order to facilitate a point capture without having to reposition it. Consequently, it is advisable to provide access for human operators to carry out various adjustments and resolve any maintenance issues. Taking these various recommendations into account, the location of the 3D scanner is indicated in Figure 5.5.

5.4.3.4. *Step 4 – communication architecture*

The fourth step consists of designing a communication architecture specific to this 3D scanner, guaranteeing both its compatibility with existing systems and the integration of new operations induced by the "Quality 4.0" module. This step, carried out in parallel with Step 3, aims to ensure harmonious integration with the MES, while maintaining the fluidity and overall production dynamics on the Industry 4.0 platform. It includes the definition of communication protocols, the choice of appropriate connectivity, the location of analysis algorithms, whether hosted locally or in the cloud, as well as data storage in the data lake. Furthermore, the cybersecurity dimension must be integrated from this phase in order to reduce risks and strengthen overall protection.

The design of a communication architecture depends both on the objectives mentioned previously, on the usage scenario and on the equipment concerned. For the usage scenario, quality is assessed by comparing the scanned surface model to the native CAD model of each part, then measuring and recording the deviations in a database. If the deviation respects the established tolerances, the part is deemed conforming to the quality repository; otherwise, it is considered defective. These measurements and analyses are performed by PolyWorks, and the results are transmitted to the MES through a procedure described in section 5.4.3.5.

Concerning the equipment, the main components are the FARO Cobalt scanner, a processing computer (PC) located nearby as a host and a virtual server hosting the MES. The OPC UA communication protocol is favored for its ability to ensure robust interoperability between different systems by standardizing communication and data mapping. Its flexibility and extensibility also facilitate the smooth integration of the FARO scanner with the MES. The MES is configured with an

OPC client to receive results from the host computer via the virtual local area network (VLAN 195), dedicated to the platform, thus guaranteeing compatibility with the overall communication architecture.

The host computer, connected to the FARO scanner by a dedicated Ethernet connection, launches PolyWorks to control the scanner and receives data in the form of a point cloud, which it transforms into a surface model. Figure 5.9 presents a schematic overview of this connectivity architecture.

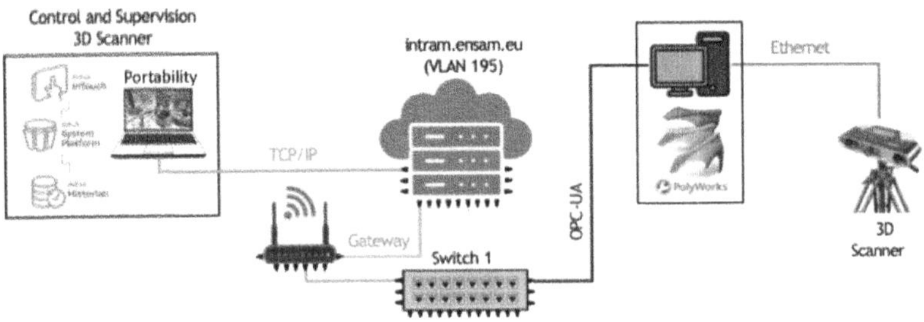

Figure 5.9. *Communication architecture*

5.4.3.5. *Step 5 – functional architecture*

During the execution of a production plan, if the MES identifies an operation dedicated to the quality control of a part, it sends a command to activate the process as soon as the robot has placed the part on the scanner table. This command activates an execution variable in the PolyWorks software, which can then execute macro scripts using its macro editor. Thanks to this script, the system can automate certain tasks such as digitization and alignment. Furthermore, another macro script has been prepared to save in a text file the results, indicated as pass or fail, as well as the deviations measured by PolyWorks. However, for each part, the quality control projects must be previously created in the PolyWorks workspace, and the characteristics must be defined in the PolyWorks Inspector module.

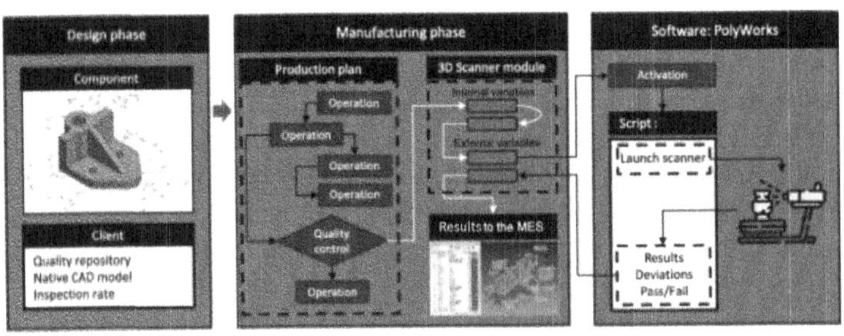

Figure 5.10. *Functional architecture*

As shown in Figure 5.10, the design process generates for each part its native CAD model as well as its quality repository, where the geometric specificities and tolerance intervals are specified. Furthermore, the customer may require a minimum inspection rate, variable according to the sensitivity of the manufactured products. For example, in the aeronautical sector, where traceability is crucial, the inspection rate is 100% of parts. Moreover, two new "robot" commands must be developed: one to place and the other to remove parts from the scanner's rotary table. Thus, the new trajectories developed for this purpose will join the library of standard trajectories, which the MES can call upon.

The manufacturing process is managed by the MES, which generates a production plan to execute each operation in a precise schedule. When the plan requires a quality control operation, the MES activates the macro script programmed in PolyWorks. During the execution of this script, the scanner is activated, the digitized data is collected, alignments are performed and finally the comparison with the CAD model produces a report on the observed deviations as well as a logical variable indicating whether the component is conforming or not with respect to its quality repository. These results are then sent to the MES and displayed in its graphical interface.

In order to achieve this objective, input/output variables are defined to facilitate the reception and sending of necessary data or information in a timely manner. For example, the variable Launch_scan is our output defined in the MES to activate the 3D scanner. However, there are also certain internal variables specified in the MES and PolyWorks. These variables are neither inputs nor outputs; their use is restricted to the software. Similarly, certain input/output variables are essential for establishing communication between equipment on the Industry 4.0 platform, particularly with the robots that must execute transfer trajectories before or after the quality control process.

To validate this methodology, an experiment is presented in section 5.4.4. It aims to apply and verify the consistency of the different phases suggested for integrating a quality device into a digitalized production system. The challenges consist not only of integrating a new industrial module but also of transforming it to make it compatible with a digitalized industrial ecosystem, particularly in terms of operability, supervision, architecture, data analysis, among other aspects.

5.4.4. *Experimentation*

On the Industry 4.0 platform, the integration of the FARO 3D scanner into the MES made it possible to connect the quality control system to the entire production process. This connection enabled the autonomous execution of quality control by the

MES. For this experiment, a part of simple geometry, a hollow three-way connector with an outside diameter of 40 mm, was specially manufactured (see Figure 5.11). Complementary operations, not detailed here, involve the use of a mobile manipulator to place and remove the part from the rotary table before and after quality control.

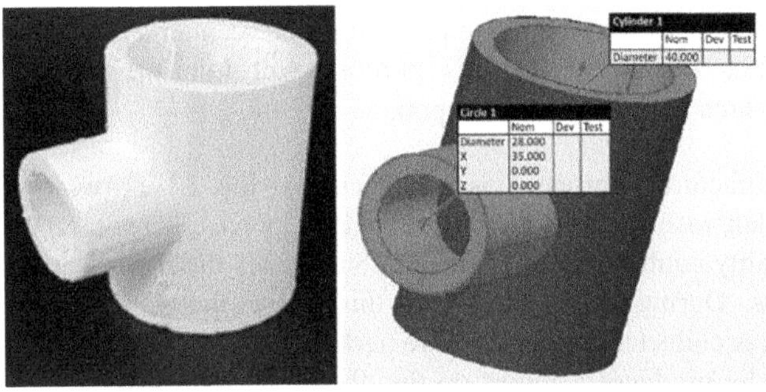

Figure 5.11. *A hollow three-way connector*

The process begins with the creation of a new project entitled "Exp 1" in the PolyWorks workspace. Next, the CAD file is imported, thus making it possible to define the geometric or directional characteristics to be inspected. To simplify the process, only the critical characteristics, namely "circle1" and "cylinder1", were selected.

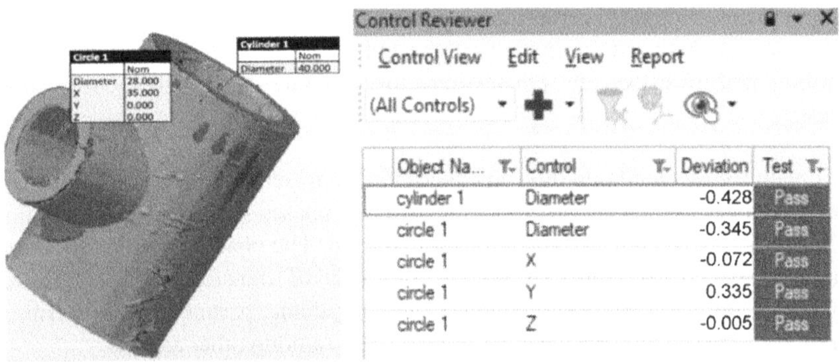

a) b)

Figure 5.12. *Alignment and results. For a color version of this figure, see www.iste.co.uk/benfriha/industrialprocesses.zip*

The digitization and quality control processes are launched by executing the quality control module, configured from the MES, according to the specified input data (number of characteristics and the project identifier). At the end of the process, the results are transmitted to the MES and saved in the database for possible subsequent analysis.

Figure 5.12(a) shows the alignment or superimposition of the model reconstructed from collected data with its CAD model. The Control Reviewer section of PolyWorks, visible in Figure 5.12(b), shows in the last column the Pass value, indicating that the measured characteristics are within the appropriate tolerances and geometric range. It also displays the deviation quantitatively.

The graphical interface of the MES, designed for this experiment, displays the quality control results, as shown in Figure 5.13. When the selected project (here "Exp 1") is entered and the scanner is activated, the feedback indicates in real time the different phases of the scanner, whether it be the reception of a command, scanning or the finalization of alignment. Finally, once the process is completely finished, the results are transmitted and can be consulted in the MES.

This experimentation of the proposed methodology demonstrated the value of adapting quality control processes to the operating mode of digitalized production lines. However, it also highlighted certain limitations, notably the impossibility of scanning highly polished surfaces as well as dark, shiny and transparent surfaces. Another challenge encountered was the noise detected during part digitization, which considerably slowed down the alignment process in the PolyWorks software. Recalibration of the scanner after a few tests made it possible to reduce this noise significantly.

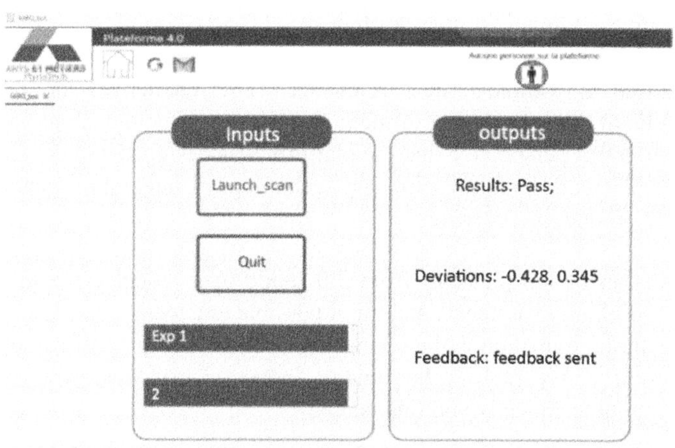

Figure 5.13. *MES interface*

Another limitation concerned communication reliability, with connection difficulties having been observed during certain experiments. To overcome this problem, the protocol used, OPC UA, was reinforced by an application developed in Python in order to improve the fluidity of exchanges.

Despite the spectacular advances in quality control methods and technologies in the era of Industry 4.0, the integration of quality systems into digitalized production systems still requires developments and research. This chapter has finally presented and experimented with a global methodology for the integration of cyber-physical modules dedicated to quality control.

5.5. Conclusion

The sciences of industrial digitalization constitute a relatively recent discipline, but their importance for the future of industry is essential. The development of this field is experiencing rapid growth. Although some companies rightly adopt a cautious approach to the transition toward Industry 4.0 technologies, it is evident that this evolution is irreversible. One of the key characteristics of Industry 4.0 is the integration of technologies from various domains, which, although complex, brings added value by offering new functionalities and improved operational performance, in the context of new societal constraints. Indeed, the balance between the search for performance at lower cost and the evolution of consumer expectations regarding decarbonization, recyclability and respect for the environment may seem contradictory.

It is within this framework that the Industry 4.0 platform plays a crucial role by providing answers through research projects, the majority of which require experimentation to validate their relevance. The challenges of research in Industry 4.0 lie at the intersection of methodological, technological and algorithmic domains. It is in this perspective that the platform was designed and developed, with the objective of creating multidisciplinary technological integration methods aimed at digitalizing production processes.

The transition of the platform from 3.0 mode to a model close to 4.0 was a particularly enriching experience, making it possible to generate new knowledge that supports the digital transition while taking into account existing infrastructures. This approach found a favorable echo among industrial partners, who contributed significantly, both technically and financially.

The level of maturity reached by the Industry 4.0 platform today constitutes a veritable showcase for ENSAM's know-how in a partnership framework. It is also used for pedagogical development, making it possible to engage reflection on teaching and learning aspects. The complexity and multidisciplinary dimension of the subject push teaching teams to revise not only the content of programs but also teaching methods.

In terms of investment, it is interesting to note that the share dedicated to engineering studies represents nearly half of the amount announced in the introduction, i.e. half a million euros. Indeed, acquiring industrial equipment such as cobots, 3D scanners, sensors, mobile robots and an MES is not enough to digitalize an industrial system. Numerous engineering and research works are necessary to make the system fully interoperable and endowed with intelligent supervision. Industrial equipment suppliers strive to transform their products into cyber-physical systems, thus enabling natural integration, both at the workshop level and at the cyber level, thanks to the digital twin. However, certain questions remain pending, particularly concerning the level of interoperability, the business model, the management of data generated by equipment, cybersecurity, etc.

6

Production Control

Within the framework of the Fourth Industrial Revolution, companies seek to integrate advanced digital technologies to benefit from new functionalities, notably the flexibility of production operations. At the heart of this flexibility are dynamic production operations control systems, which fully exploit the advantages of digitalization of manufacturing processes. These systems rely on technologies such as artificial intelligence (AI), advanced data analysis and real-time connectivity, enabling faster decision-making in terms of operations scheduling.

To achieve this, researchers develop real-time control algorithms that monitor and supervise production processes, detect anomalies and adjust, in certain cases, machine parameters to maintain optimal performance. For example, given the proportionality between a robot's speed and its energy consumption, it is possible to define a speed that maximizes flow while minimizing energy consumption. Furthermore, other factors such as reducing the carbon footprint can be taken into account in this optimization process.

However, the autonomy of control systems remains limited due to the complexity of computational models, the large number of variables and parameters involved, the diversity of possible scenarios and the additional constraints imposed by robots on mobile platforms. Furthermore, the legal responsibility of autonomous systems is not clearly defined, particularly in the event of decisions causing damage or accidents involving humans. Numerous research works endeavor to address this issue, which poses real challenges to scientists, confronted with multi-constraint, multi-variable and multi-objective modeling approaches.

At the same time, intelligent control systems anticipate resource, material and production capacity needs, thus optimizing operations orchestration, reducing waiting times and maximizing asset utilization. This makes it possible to gain agility and responsiveness while ensuring the efficient use of resources.

Despite the numerous advantages offered by these advanced control systems to support digitalized production processes, their implementation requires a global strategy, close collaboration between operational and IT teams, as well as a cultural adaptation to integrate these technologies into existing work processes. The effectiveness of control systems relies both on a clear functional architecture, which will be detailed later, and on the communication architecture, already addressed in Chapter 4. Dynamic scheduling requires several prerequisites, notably real-time data feedback, computational capacities for complex analyses, as well as the ability to activate actions at the level of operations, machines, quality control, assembly, storage, and more particularly, robots ensuring transfers between production units.

6.1. Functional architecture

Starting from expectations in terms of functionalities, performance and control, the design process of a global functional architecture can be initiated. Figure 6.1 presents an example of a functional architecture for the Industry 4.0 platform, which encompasses both the design phase and the "methods" phase, where all operations are prepared upstream. Thus, it is organized into five spaces interconnected by a bidirectional information flow and open to the cloud. This approach ensures digital continuity throughout the design, methods and production phases, with the possibility of extending it to the entire lifecycle of an industrial product, thus making it possible to create dynamic interaction models between these different stages.

6.1.1. *Digital continuity*

Thanks to data collected from the different phases of the lifecycle, artificial intelligence tools and computational resources offered by cloud computing, it is now possible to model dynamic links between a component of a product in the design phase and the information essential to its manufacturing, such as production costs and lead times. Thus, designers will have tools enabling them to predict costs, durations or other metrics related to the production of designed components. This will make it possible to optimize design in order to meet as closely as possible the requirements defined in the form of objective functions, such as minimizing energy consumption during the manufacturing of a component.

Furthermore, a functional architecture integrated with all phases of a product's lifecycle makes it possible to assess the distribution of its carbon footprint. Thus, digital continuity, associated with complex algorithms, facilitates the development of models for predicting the carbon profile of a product over its entire lifespan. This constitutes a valuable source of information for the design phase, particularly concerning technological choices, materials and processes.

6.1.2. *Business flows*

As mentioned at the beginning of this section, the functional architecture of the Industry 4.0 platform is organized into five spaces interconnected by an information flow organized in a "business" logic. The first space is reserved for design processes, which must integrate a certain variability at the level of a few product parameters. Digital continuity between the design phase and the industrialization and production phases will enable simultaneous global optimization of the product and its manufacturing process. The idea behind this approach is based on the fact that data generated in the production phase has the possibility of evolving a few design parameters with a view to maximizing or minimizing a single-objective or multi-objective function. Take the example of an objective function that aims to minimize energy consumption in the machining phase. To this end, optimization can focus on the variation of cutting speeds, as well as on the geometric variation of the component, so as to minimize its passage on an energy-intensive machine. Digital continuity between design, industrialization and production can generate a very promising global optimization context.

The second space is dedicated to the importation and recording of standard parametric operations. Basic operations are generated by specialized skills using specific tools. Certain operations must be encapsulated in order to make them compatible with the platform's control and command system. Take the example of the transfer function, ensured by 33 trajectories that cover almost all the platform's needs. These trajectories were programmed with parametric attributes by specialists to constitute a library of standard trajectories, which the manufacturing execution system (MES) can execute according to needs expressed by real-time dynamic control. All basic operations must be validated, imported and listed in this space before moving on to the generation of an initial production plan.

The third space (in blue in Figure 6.1) comprises two new specific functionalities, essential for the orchestration of operations in a digitalized production context, supervised by an MES. The first functionality makes it possible to design a master production plan from different predefined modules. This initial plan proposes a scheduling based solely on precedence constraints between the different operations, without taking into account configuration constraints or the load plan of the Industry 4.0 platform. Precedence constraints mean that an operation can only begin when other operations have been completed because some depend on previous ones. The objective of this approach is to maintain the possibility of executing part of the production plan on other distant and interconnected platforms. However, although this initial plan is not directly executable, it constitutes an essential prerequisite for the second functionality.

The second functionality makes it possible to launch and supervise a production plan. From its initial version, this functionality generates a plan adapted to the workshop configuration, including the number and layout of machines, fixed robots, the quality control device, mobile robots, as well as material flows. The same applies to the workshop load plan, which consists of determining the rate and periods of occupation of the various equipment in order to integrate the new production plan. The production plan thus generated is executable because the calculated scheduling takes into account precedence constraints, the configuration and the workshop load plan. In this version, the MES assigns standard values to production parameters, such as robot speed or cutting conditions for machining machines. It is, however, possible to activate complex multi-objective optimization algorithms, which will adjust these parameters according to priorities defined by production supervisors, in order to refine the execution of the plan.

In the production phase, operations control in a context of digitalization of processes, procedures and flows is dynamic. It consists of supervising the execution of all production sequences and reacting in real time in an appropriate manner to disturbances or incidents. Thus, the MES is endowed with intelligence and a certain autonomy to adjust not only the scheduling of operations but also the values assigned to production parameters.

For example, in the event of an unexpected machine stoppage, the flow will be temporarily redirected to other equivalent machines, while analyzing the malfunction and intervening for repair. This situation requires generating a new scheduling by excluding the faulty machine from the layout of available machines, which underlines the importance of having sufficient computational resources so that the MES can regenerate schedules as often as necessary in response to quality variations, breakdowns, priorities or errors. Finally, the data collection and analysis system reports information to the MES, some of which is transformed into performance indicators viewable through the digital twin.

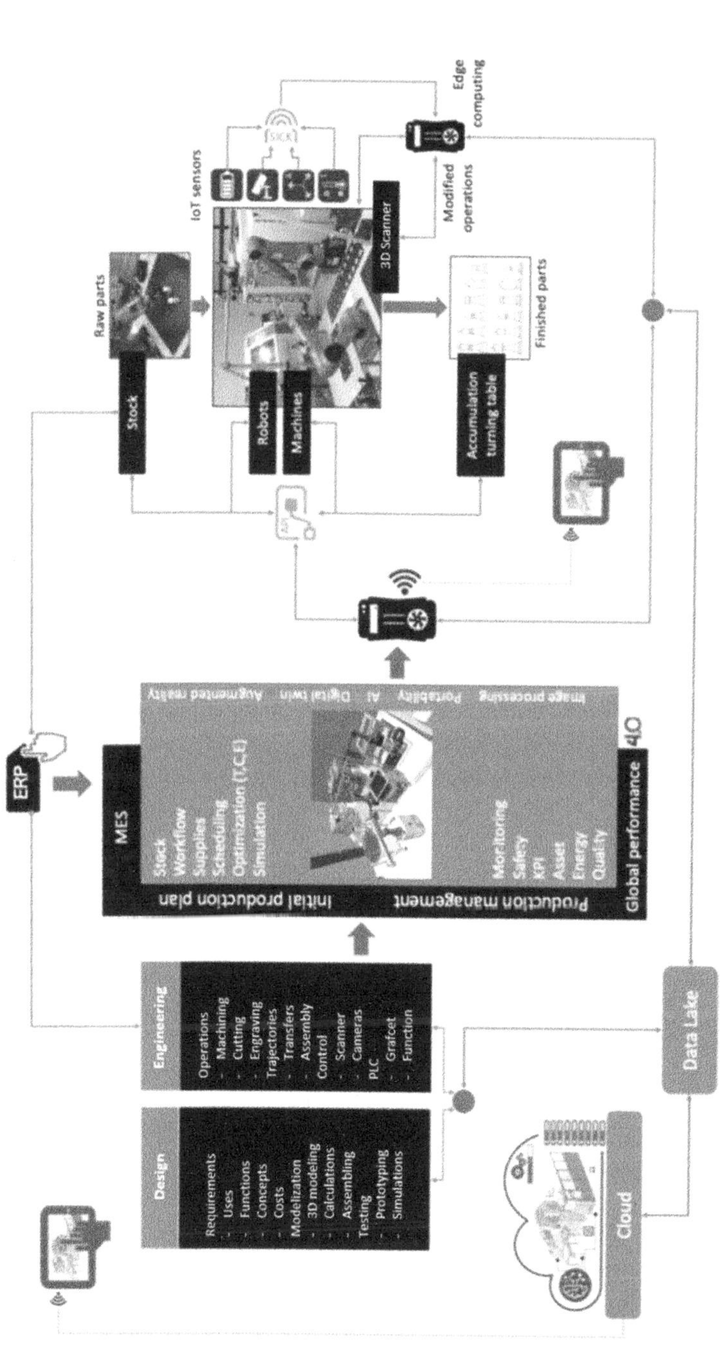

Figure 6.1. *Architecture of the industry 4.0 platform. For a color version of this figure, see www.iste.co.uk/benfriha/industrialprocesses.zip*

The right part presents a simplified diagram of the Industry 4.0 platform, characterized by production machines and equipment, storage areas, as well as industrial Internet of Things (IIoT). IIoT systems are connected to the local computer. The mass of data generated during the design, industrialization and production phases is centralized in a data lake, with a gateway to the cloud. This functional architecture guarantees digital continuity and makes it possible to valorize data from design, industrialization and production, with a view to simultaneous optimization of the product and the production process.

6.2. Control systems

A production supervision software, or MES, is a computer system designed to connect, monitor and supervise manufacturing processes and complex data flows at the workshop level. Its main objective is to guarantee the efficient execution of production operations and to improve the overall performance of industrial activities.

The MES collects and tracks precise data in real time throughout the production lifecycle, from the manufacturing order to the delivery of the finished product. For each product, it captures information relating to genealogy, performance, traceability, material management and work in progress, while monitoring other ongoing activities in the factory. This data offers decision-makers a detailed view of production parameters on the shop floor, thus facilitating process optimization.

In 1997, MESA International (Manufacturing Enterprise Solutions Association) defined the "MESA-11" model (MESA International 1997), which identifies 11 essential functions of an MES. These functions include resource management, operations scheduling, document tracking, product traceability, quality management, production process monitoring, maintenance management, performance management, material management, human operations tracking and data analysis to support decision-making.

The MES integrates into an industrial architecture standardized by ANSI/ISA-95, which merges the MESA-11 model with the pyramid architecture (Purdue Reference Model) introduced by Purdue University in the 1990s. Furthermore, this model has been widely adopted, and some researchers, such as Bertolini and Guelfo (2020), have used it as a reference to design an architecture dedicated to intelligent processes. In the pyramid architecture, the MES is placed at level 3 of the functional hierarchy, between enterprise resource planning (ERP, level 4) and process control (levels 1 and 2), as shown in Figure 6.2. This key position allows the MES to act as a crucial link between strategic management and real-time operations.

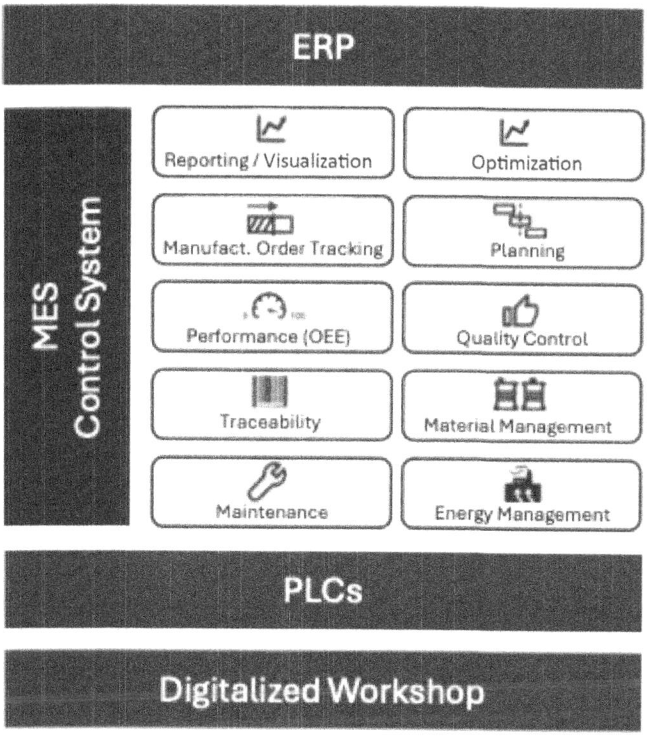

Figure 6.2. *Position of the control system*

The evolution toward Industry 4.0 has led to adjustments of the initial MESA-11 model. New frameworks have emerged to meet the growing needs for IT/OT system integration, interoperability and real-time control. A more recent model, developed by MESA International (MESA International 2016), is the Manufacturing Operations Management (MOM) framework. This integrated approach aims to optimize the management of manufacturing operations by transcending the traditional MES. It integrates essential strategic dimensions, such as global production, quality, inventory and maintenance management, while ensuring fluid interaction with ERP systems.

New trends highlight the vertical and horizontal integration of systems to connect the entire value chain, as well as real-time data management for faster decision-making. The use of artificial intelligence and machine learning, also called "intelligent computing", makes it possible to optimize production processes. Finally, system flexibility and modularity are becoming essential to adapt to new market requirements.

The MOM model addresses the challenges of Industry 4.0 and relies on new technologies such as the Internet of Things (IoT) and artificial intelligence. The adoption of an MES in industrial systems of Industry 4.0 is no longer an option but a necessity. The growing complexity of digitalized production systems, combined with the integration of technologies such as IoT, autonomous robots and artificial intelligence, requires dynamic planning and real-time control. This was confirmed by MESA International, which published in 2022 a new model for smart manufacturing (MESA International 2022). Furthermore, the MES plays an essential role in centralizing data management, guaranteeing process traceability and optimizing workshop performance. Without an MES, companies risk losing efficiency, failing to meet flexibility requirements and missing the opportunities for continuous improvement offered by these intelligent technologies.

In the following sections, the key concepts of a control system will be examined in detail, accompanied by an experiment on the development of an MES for the Industry 4.0 platform.

6.2.1. *Planning*

The production planning phase is a crucial technical aspect that requires, on the one hand, knowing the production capacity of machines and, on the other hand, knowing the production programs previously defined in terms of quantities, specificities and lead times. Added to this are other elements, such as resource availability, supplies, scheduled stoppages for maintenance, as well as machine and programmable logic controller (PLC) programming, all of which must be taken into account. Traditional industrial planning methods no longer meet the new requirements for flexibility, responsiveness and optimization necessary to adapt to new internal or external constraints on the company.

In contrast, digitalization confers upon the production system increased agility, directly resulting from the distributed architecture, the cyber-physical components of the production line, the modularity of operations, digital interoperability and robots ensuring the physical flows of parts within the workshop and with its environment. This configuration opens up new perspectives for the workshop. It enables the implementation of dynamic scheduling of production operations. Furthermore, it facilitates the reorganization of production lines to adjust or increase production capacity.

The concept of modularity of operations is based on the encapsulation of basic operations into standardized parametric modules, thus making each operation compatible with a digitalized production environment and increasing the flexibility of flows in real time. For example, in a conventional approach, G-code is loaded and activated manually by an operator, after customary verifications. In contrast, in a digitalized environment, G-code is integrated into a module that interacts with the machine's digital system and the control system. This module makes it possible to autonomously verify the activation conditions of the G-code, such as the presence of the part in the machine or the closing of the machine door.

Furthermore, this module includes algorithms capable of executing actions ordered by the MES, such as adjusting process parameters in real time, or acting according to its own analysis of surrounding data. For example, an algorithm can be dedicated to monitoring the progress of the operation and detecting possible anomalies, while another can focus on optimizing manufacturing conditions, such as speeds or depths of cut, necessary to achieve the required geometric and dimensional characteristics. This approach avoids reworking parts and contributes to the integrated optimization of the production process.

6.2.2. *Modularity*

As previously mentioned, modularity contributes to the transition from linear planning to dynamic scheduling capable of adapting to evolving circumstances and requirements, without requiring sophisticated reprogramming work. Furthermore, the standardization of modules enables the creation of libraries of reusable modules by the MES.

Thus, thanks to this modular approach, designing a flexible production plan consists of assembling, for each product to be manufactured, specific modules and standard modules whose parameters must be defined. A specific module can contain the G-code, while a standard module can contain a standard trajectory used for the physical transfer of parts between two known positions.

The parametric dimension expresses a variability that allows modules to adapt to the context. Modules can be defined in different ways, depending on the desired standardization. However, six categories emerge as representative of the most common operations, summarized in the first column of Table 6.1.

Modules	Examples of missions	Equipment
Transfer	Transfer between stocks and machines	Kuka Agilus 1,100
Manufacturing	Machining of cylindrical parts	Tormach Lathe
	Engraving of parts	Universal Laser
Quality	Dimensional quality measurement	Faro 3D Scanner
Control	Inventory status recording	Fixed Cameras
	Verification of parts in machines	Embedded Cameras
Assembly	Assembly with one robot	Kuka Agilus 900
	Assembly with two robots	Robots kuka 900 and 1,100
Movement	Robot movement without parts	Kuka Agilus 1,100 and 110

Table 6.1. *Standardized production modules*

6.2.3. *Production plan*

In the era of digitalization, production plans have undergone a radical transformation. Thanks to the integration of digital technologies, the way companies design, manage and execute their production plans has been profoundly modified. Connectivity and IoT systems have revolutionized the monitoring and control of production processes. Thus, production status and performance are tracked almost in real time.

Production planning and management systems are now capable of analyzing vast quantities of data in real time, which enables faster and more accurate decision-making. Certain digital platforms dedicated to industrial digitalization offer an environment for developing production planning and monitoring applications, such as MES. This work is far from simple; it requires specialized multidisciplinary skills, a deep knowledge of the production environment and the ability to properly characterize the targeted functionalities.

In summary, the era of digitalization has brought significant improvements to production plans, transforming once rigid and linear processes into dynamic, flexible and highly optimizable systems. This transformation offers increased personalization, as well as improved traceability and transparency in the production chain. Digital systems make it possible to track and act on each step of the manufacturing process, from raw material to finished product. However, this new production dynamic requires advanced algorithmic tools, considerable computational resources and specific skills to methodologically and scientifically support this evolution.

To this end, the design of a flexible production plan adapted to digital production lines is organized into two phases. The first, called the "initial" phase, is a first projection that takes into account precedence constraints between modules. These are first-level constraints that define the order in which certain operations must be carried out. These constraints guarantee that the stages of the production process are executed in a logical sequence, thus ensuring the feasibility of the final product. For example, tapping operations cannot be executed before drilling operations. Algorithms such as linear programming, the critical path method (CPM), and program evaluation and review technique (PERT) can be used to generate production plans by integrating precedence constraints. Other constraints, such as workshop capacity in terms of machines, resources or storage, can be added to define all dependencies between the different operations. For example, if two operations can be executed simultaneously on the same type of machine, but the workshop only has one machine available, then the initial plan will systematically propose scheduling these two operations sequentially.

This initial plan cannot be executed as is, but it offers a global overview of lead times, flows and supplies. Similarly, it can prove useful if we consider distributing the manufacturing of a product between different digitalized and connected factories. In this perspective, the components of a product could be manufactured in various distant and interconnected production units. The organization of production would then no longer be limited to the capacity and constraints of the Industry 4.0 platform but would extend to other digitalized production sites. For this, MES must collaborate to optimize resource utilization by sharing production schedules, machine capacities and raw material inventories. The interoperability and interconnection of control systems make it possible to coordinate outsourced operations and define production priorities, while ensuring dynamic updating of the progress of multi-site production.

The second version of the production plan, called the "optimized" version, is generated by the MES by proposing several possible scheduling and flow scenarios, adjustable according to the weighting of the retained optimization criteria, such as cost, speed or energy. Optimization can be single-objective or multi-objective, aiming, for example, to generate a production plan offering the best production time at the lowest cost. Furthermore, the models used for energy prediction of each operation can evolve during the execution of manufacturing sequences, particularly if they use learning approaches. Consequently, the optimization will be updated, leading to modifications of certain parameters of the concerned modules. Once the plan is optimized, validated and saved, the MES will be able to schedule its execution.

It is in this context that the third version of the production plan intervenes, by generating an operations schedule under the constraints of the current load plan in the workshop. Indeed, available production data, transmitted in real time to the MES, makes it possible to schedule the execution of all modules of the production plan according to resource availability in the workshop and priorities set by the company. Thus, certain modules, whose manufacturing is possible, will be executed totally or partially, while others will be placed in a queue. The modularity of operations allows the MES to dynamically schedule operations and adjust parameter values according to the evolution of the context and priorities. For example, the MES can reduce speeds to minimize the energy consumption of operations, but this parameter can be adjusted upward in the event of unforeseen events likely to impact lead times. Thus, MES priorities are generally compromises between reducing energy consumption, meeting production deadlines, and minimizing costs. Finally, the control of the production plan in dynamic mode will be addressed in section 6.2.4.

6.2.4. *Control*

Production control aims to guarantee the execution of manufacturing operations, whether successive or simultaneous, while respecting customer quality and quantity requirements. This process takes into account both manufacturing modules and necessary logistics, including tooling, raw materials, maintenance, as well as unforeseen events that may occur during production. Industrialization, which constitutes an essential preparatory phase upstream of production, aims to define the set of operations grouped by module, which will be integrated into a digitalized production plan. In this context, the MES is distinguished by several key characteristics.

First, it is capable of designing a production plan by chronologically assembling standard and specific operations, whether sequentially or simultaneously, while taking into account the various production constraints and priorities. Furthermore, the conditions for transitioning from one operation to the next are also determined, often according to quality criteria or the availability of the mobile robots necessary for component handling. Finally, the MES must be able to supervise production dynamically, relying on advanced algorithms and real-time upstream data, which enables optimal responsiveness to various variations.

Figure 6.3 summarizes six interfaces on which an MES relies. The *first interface* is dedicated to the supply of raw materials or components necessary for the assembly of the finished product. This function is relatively conventional in its operation. The chain of operations necessary for the acquisition, management and renewal of material stock is managed by the ERP system. Stock status is monitored in real time thanks to the digitization of output flows. Each stock output or scrapping movement is recorded and tracked electronically, enabling instant and accurate updating of available stock levels. Furthermore, this makes it possible to better anticipate replenishment needs and optimize surplus management.

The *second interface*, entitled "operations", is reserved for defining production operations specific to the products to be manufactured. The implementation of specific operations in the MES is organized in two steps. The first step takes place on systems, such as computer-aided manufacturing (CAM) tools dedicated to generating tool paths for computer numerical control (CNC) machining machines. Indeed, all specific operations necessary for a given product must be validated upstream. Then comes the second step, which consists of preparing their integration into the MES by means of an encapsulation action in order to make them compatible with their digital environment.

This process relies on advanced technical knowledge of the Industry 4.0 platform and follows a precise codification in order to be identifiable in different interfaces. Furthermore, production operations that involve robot trajectories are defined beforehand in terms of encapsulation, standardization and parameterization. These definitions are then recorded in the library of standard operations.

In this perspective, an in-depth analysis of possible robot trajectories on the Industry 4.0 platform led to defining 33 trajectory sequences. The input parameters necessary for activating these operations include speed, starting point, gripper, end point, etc. These sequences can be combined to meet all transfer needs in the different accessible workspaces.

Table 6.2 presents, in the second column, an example of standardized production operations, capable of being integrated into various production programs. The objective is to identify and define all operations necessary for manufacturing, inspections, transfers, assembly, etc., thus guaranteeing the production of a component, from the initial raw material stock to the delivery of the manufactured part, inspected and positioned at the accumulation table for removal.

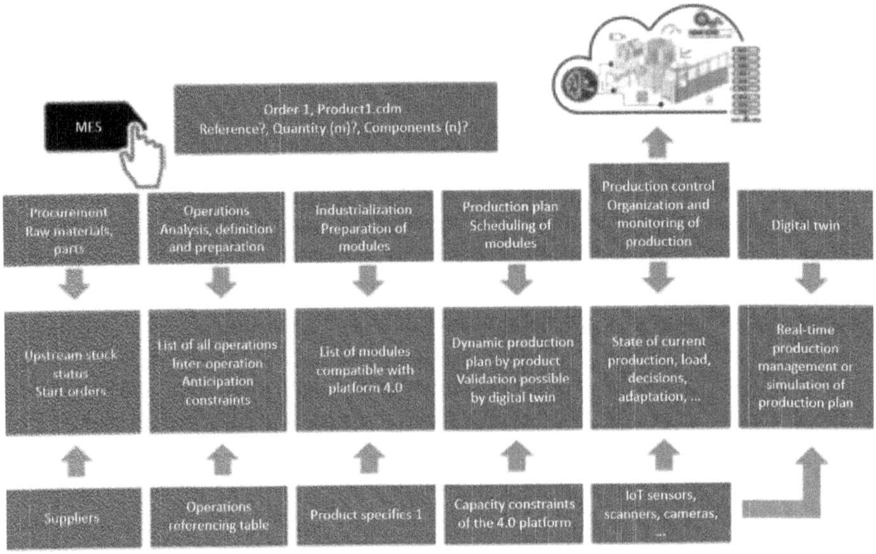

Figure 6.3. *MES organization chart of the Industry 4.0 platform*

Certain operations may require specific parameterization, such as the frequency at which dimensional or geometric quality control must be performed by the 3D scanner. This parameterization can be defined as a percentage relative to the total number of parts to be produced.

Furthermore, if the MES detects a dimensional deviation, the inspection frequency could be proportional to the acceleration of this deviation. This approach makes it possible to optimize the use of the 3D scanner, which requires computational resources and additional time.

Similarly, the stroke of the grippers used by robots to grasp parts is another parameter whose variability contributes to smoothing part transfers. Indeed, the typology of the various manipulated components suggests maintaining the "part grasping" aspect as an input parameter for transfer modules.

Machines	Operation code	Definition
Robot K900	W_RF_Pick_InitStore	Retrieve raw part from the initial stock
Robot K900	W_RF_Drop_TransStrore	Place raw part in the intermediate stock
Embedded Camera	W_RM_Control_Tour	Check the presence of raw part in the lathe
Tormach Lathe	Prog_Tour_1	Start machining program
Robot K1100	W_RM_Pick_Tour	Retrieve machined/manufactured part
Robot K1100	W_RM_Drop_Scanner	Place part to be scanned
Scanner FARO	Prog_Frao_1	Start scanner
Robot K1100	W_RM_Pick_Scanner	Retrieve part from a scanner
Robot K1100	W_RM_Drop_TransStore	Place part in the intermediate stock
Robot K900	W_RF_Pick_TransStore	Retrieve part from the intermediate stock
Robot K900	W_RF_Drop_Conv	Place part on a conveyor

Table 6.2. *Examples of production operations*

The *third interface*, entitled "industrialization", is dedicated to the preparation of modules, where each module is composed of a batch of operations from the previous phase, organized in a precise order. These modules can represent a logical production phase or be specifically dedicated to a particular product. Each module retrieves the parametric attributes of the operations that compose it and must be simulable and executable. A relevant example is the recurring dimensional inspection of a component using the 3D scanner. For this, the quality control phase is orchestrated by a standard module integrating several operations: the transfer of the component to the scanner's rotary table, its activation, the retrieval of the point cloud, analysis, and finally, the reintegration of the component into the production flow if it is conforming. In the event of non-conformity, a direct output is envisaged.

Modules	Operation description	Definitions
Raw part transfer from initial stock to the "lathe" machine	W_RF_Pick_InitStore ↳W_RF_Drop_TransStrore ↳W_RM_Pick_TransStore ↳W_RM_Drop_Tour	This module transfers a raw part from the initial stock to the lathe chuck
Presence check	W_RM_Control_Tour	In situ check of the raw part presence
Machining on lathe – "Pion" part	Prog_Tour_1	Start machining program
Case 1 **Transfer of the machined part from chuck to conveyor without quality control**	W_RM_Pick_Tour ↳W_RM_Drop_TransStore ↳W_RF_Pick_TransStore ↳W_RF_Drop_Conv	Retrieve the machined part from chuck and evacuate it to the accumulation table
Case 2 **Transfer of the machined part from chuck to conveyor with quality control**	W_RM_Pick_Tour ↳W_RM_Drop_Scanner ↳Prog_Frao_1 ↳W_RM_Pick_Scanner ↳ W_RM_Drop_TransStore ↳ W_RF_Pick_TransStore ↳W_RF_Drop_Conv	Retrieve the machined part from chuck, place the part in the 3D scanner, then evacuate the part to the accumulation table

Table 6.3. *Examples of production modules*

This interface plays a crucial role by allowing engineers to prepare production modules without encountering issues of interoperability, communication, robot and PLC programming, thus offering a significant advantage in terms of productivity.

Several approaches can be considered for defining modules from the operations identified in Interface 2. The first consists of grouping a minimal set of operations under a clear business identification accessible to all actors involved in production. Table 6.3 illustrates this approach by showing that some modules group a single operation, while others integrate several. For example, the "lathe machining – pin" module comprises the single operation "Prog_Tour_1" specifically dedicated to machining a particular component. This choice underscores the importance of the operation and the necessity of making it visible in the production plan. Similarly, the "presence check" module for the raw part in the lathe chuck consists of a single operation, due to its frequency of use, which is desynchronized relative to the production cadence of the concerned component. As a general rule, module configuration relies on the experience of engineers, who anticipate production constraints and productivity requirements in order to optimize operational efficiency.

The second approach aims to group a maximum number of operations in order to make the production plan more explicit. However, this can also lead to an increase in the number of modules necessary to adapt to different situations. Table 6.3 presents two examples, Case 1 and Case 2, illustrating the transfer of the machined part from the chuck to the accumulation table by mobilizing the two robots. These two cases are distinguished by the inclusion, in one of them, of a "quality control"-type operation of the machined part, performed by the 3D scanner, before its transfer to the accumulation table. By grouping operations into modules, their visibility and clarity for users of MES are considerably improved.

The *fourth interface*, entitled "production plan", is dedicated to the finalization of a production plan for a given product. The production engineer retrieves all modules necessary for the execution of a specific manufacturing order and imposes precedence constraints on some of them. Furthermore, certain modules are parameterized, notably quality control modules, which must be activated proportionally to the number of manufactured components. Other modules, including robot trajectories, integrate parameters such as the spatial positions where the part must be retrieved and then placed, as well as the speed at which the robot must travel these trajectories. These input parameters can be characterized by real, logical or integer variables and are framed by minimum and maximum values for some of them. In this regard, Figure 6.4 shows the associated parameters necessary for the execution of each module.

In this phase, the MES offers several options. The first option consists of taking into account the imposed precedence constraints and suggesting all possible scheduling scenarios of production modules for a given product. Figure 6.4 shows an example of a production plan for the manufacturing of a pin, organized into several possible flow scenarios, two of which are particularly relevant. The first flow transfers the machined part from the lathe chuck to the 3D scanner, activates the scanner in order to measure and record dimensional deviations, then transfers it from the scanner to the conveyor. The second flow directly transfers the machined part from the lathe chuck to the conveyor without going through the 3D scanner. The alternation between these two flows is conditioned by the desired dimensional inspection frequency for a population of machined parts. This parameter, defined by the MES during the control phase, can be fixed or varied proportionally to the evolution of dimensional deviations observed over time.

This alternation is made possible thanks to the integration of functions that condition the transition from one operation to the next. These functions, essential for the execution of the production plan, bring operational autonomy and can integrate decision-support applications based on mathematical or predictive models. In the example presented, they are integrated manually, but it is entirely conceivable that these functions could be suggested by an intelligent algorithm capable of interpreting operational flexibility needs and proposing realistic and evolving configurations. Furthermore, dynamic scheduling requires the possibility of reconfiguring in real time all modules in the course of execution, and adjusting or modifying parameter values.

Moreover, the recording of dimensional deviations of machined parts relative to their CAD model constitutes a real opportunity for optimization to maintain an optimal production rhythm. This dynamic quality control makes it possible not only to reduce interruptions related to settings and adjustments but also to minimize the number of rejected parts due to dimensional and geometric non-conformities.

Concretely, two actions can be integrated and supervised by the MES in real time. The first consists of readjusting the "depth of cut" parameter at the level of the "machining operation" module, which makes it possible to compensate for the measured deviation and to get closer to the nominal value of the dimension. However, this action proves complex because G-code programs are inviolable and cannot be modified during execution. To circumvent this difficulty, one solution consists of decomposing the manufacturing sequence into several G-codes, thus facilitating the adjustment and regeneration of the last G-code according to the measured dimensional deviations, without requiring the removal of the part from the chuck. However, this approach increases the complexity of operations and manufacturing modules. The second action concerns the spacing of the inspection frequency of machined parts, particularly if the deviation is relatively stable or evolves in a limited manner. This approach makes it possible to reduce the use of the mobile robot and the 3D scanner for the manufacturing of the "pin" component, while offering the possibility of reallocating them to the production of other components.

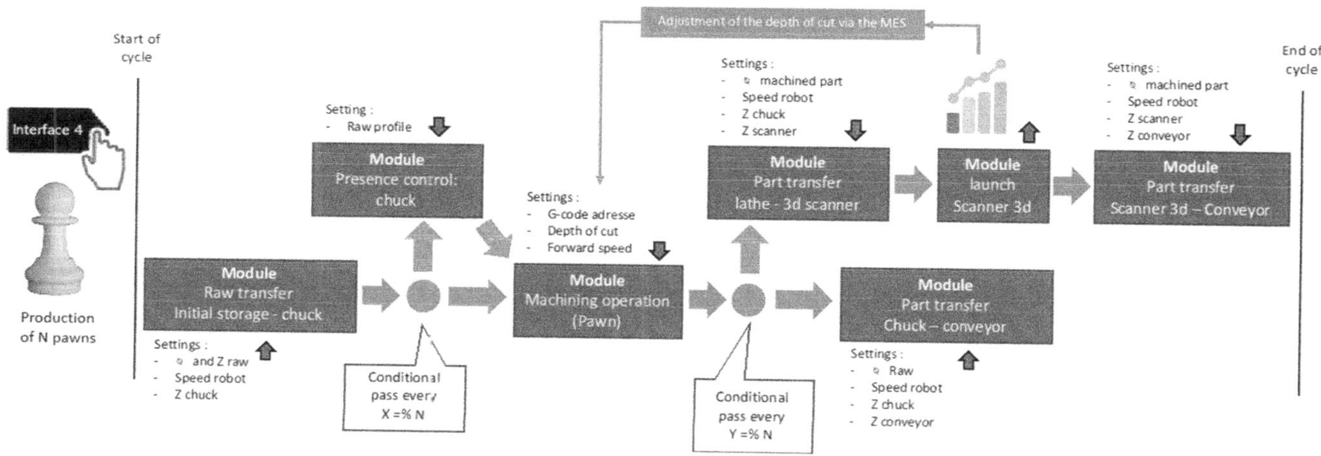

Figure 6.4. *Example of a production plan*

The second option proposed by the MES aims to optimize the results obtained from the first option by comparing them with optimization criteria and objective functions. These criteria may include minimizing the execution time of the production plan, reducing costs or decreasing energy consumption. Optimization can focus on a single criterion, qualified as single objective, or on the combination of several criteria known as "multi-objective". For example, this could involve seeking optimal execution times while limiting associated energy costs.

To do this, the numerical methods used vary module parameters such as depth of cut, feed speed, robot speeds, etc., and measure the targeted performance at output in order to suggest values enabling optimization of the production plan. The variability and complexity of the production system condition the choice of numerical methods to be used, ranging from simple decision trees to genetic algorithms, or even deep learning modeling methods.

The results obtained after optimization calculations are considered as non-contextualized references with respect to the dynamic constraints of the platform. In the "control" phase, the MES can evolve them to make them compatible with the load plan in execution. At this stage, the different production plan scenarios can be simulated using the digital twin, which will be mentioned in "Interface 6".

The *fifth interface*, entitled "control", prepares the launch of a production plan and ensures real-time control. In this interface, the initial production plan, selected and validated during the previous phase, is loaded and positioned for its execution. The MES manages the synchronization of incoming production plans with those in execution. During the preparation of the launch of a new production plan, the MES recalculates a new global schedule taking into account the priorities of each plan, the workshop load plan, scheduled stoppages, equipment status and safety instructions.

In this active phase, the MES has a level of autonomy and must therefore be capable of launching production. To do this, the MES uses complex computational applications to determine the precise moments to launch or temporarily suspend module execution, define optimal parameters for each module, control the progress of the different phases of each operation, analyze reports and manage mobile robots in terms of circulation, battery charging, occupation of functional spaces and the actions they can execute, particularly for those equipped with manipulator arms and endowed with a storage space. Furthermore, the MES records the movements of components in incoming, outgoing and intermediate stocks.

Moreover, the variation of manufacturing parameters, essential for multi-objective optimization, increases the complexity of dynamic flow management. This new context redefines the roles of managers, production operators, preparers, data analysts and advanced control systems. Production managers must establish priorities concerning parameters to be optimized, such as time, cost or energy consumption. Operators intervene mainly for maintenance or in the event of unforeseen events. Preparers define the production files for each component, including G-codes, transfer parameters, the quality repository, assembly ranges, etc. Data scientists analyze data, identify phenomenological correlations between the production plan, the behavior of the production system and their impacts on produced parts. They also develop predictive models, thus making MES more intelligent and autonomous.

The *sixth interface*, entitled "digital twin", can be activated from Interfaces 4 or 5. During the preparation phase (Interface 4), it makes it possible to simulate in immersive mode the progress of a production plan without disrupting ongoing operations. This 3D simulation plays an essential role: it validates the different organizational configurations and verifies processes before their implementation. Furthermore, it anticipates and prevents potential malfunctions such as collisions, inconsistencies, performance losses or flow interruptions. The digital twin, developed in Chapter 7, thus proves to be a powerful decision-support tool, facilitating collaboration between the different actors in production.

In the control phase (Interface 5), the digital twin fully reveals its potential. It makes it possible to visualize in real time the progress of the process from upstream data and enables actions to be triggered on the production chain, thus creating a bidirectional effect. Its portability facilitates the sharing and management of data flows. Information from physical objects in the workshop, transmitted via the MES, is recontextualized and updates the digital twin, reflecting the real execution conditions of production plans. The 3D scene offers an interactive visualization and displays performance indicators such as cadence, machine status, overall equipment effectiveness (OEE) or stoppages. This multi-scale and multi-view display is accessible to the various stakeholders, whether present locally or remotely.

This system opens up new perspectives for telecollaboration, notably in fields such as failure analysis, robot trajectory calculation, predictive maintenance, functional block programming and many others. These sectors are often those where companies lack in-house experts. Concerning training, the digital twin allows production line operators to acquire skills safely and effectively. Thanks to realistic simulations, it offers a risk-free environment in which employees can practice on virtual machines, understand complex processes and perfect their skills before production launch. This technology enables personalization of training programs and continuous updating of skills, according to industry evolutions.

In conclusion, MES play an essential role in the orchestration and optimization of manufacturing processes. They integrate advanced technologies and offer expanded decision-making autonomy. However, their design presents significant challenges. The need for harmonious integration with varied equipment, existing software and complex data flows requires a methodical and personalized approach. Furthermore, growing requirements in terms of flexibility, responsiveness and traceability further complicate MES development. Companies must therefore reconcile their performance objectives with operational realities in order to design systems adapted to their specific needs.

7

Digital Twins

The digital twin (DT) is establishing itself as a revolutionary tool. This concept, although it lacks a universally recognized definition, is perceived as a digital replica of a physical system. Some researchers describe it as a dynamic entity evolving in parallel with its physical counterpart, thanks to bidirectional interaction. The DT can be exploited in real time for the supervision of production processes through data analysis or in deferred time for simulation and verification needs. This tool offers new perspectives for interacting with advanced production systems by providing an integrated vision that transcends the traditional boundaries between the digital and the physical.

The adoption of DTs in digitalized industrial systems is becoming a necessity. It transforms production, increases performance and reduces costs, notably through its ability to simulate, test and optimize industrial processes in a virtual and shared environment before their actual deployment. Furthermore, recent DTs can embed intelligent applications, which exploit collected data, in order to increase their capacity to control and act dynamically in the event of unforeseen events or deviations in process conformity, potentially impacting the quality of manufactured parts.

Currently, this technology is in full evolution and is establishing itself as an essential pillar of Industry 4.0. However, the concrete implementation of DTs encounters major obstacles, due to several factors such as the lack of skills, methodological gaps, limited technological interoperability and a deficit of trust towards digital platform providers. This chapter proposes a global strategy for the design and deployment of DTs dedicated to digitalized production systems.

7.1. Principles

The notion of the DT was first mentioned under the term "mirror world space" by the American researcher Grieves in 2005 (Grieves 2005). He envisioned connecting a physical space to a virtual space through a continuous flow of information and data. However, due to the complexity of the concept and the technological limitations of the time, the idea was not immediately adopted.

The concept was brought back into fashion a few years later, this time by NASA, which renamed it "Digital Twin". The space agency envisioned using the DT for the maintenance of its space equipment. This concept finds its origin in the early stages of space exploration, where a backup physical model of the shuttles was kept on Earth to help technicians with problem-solving during missions. Unlike the creation of a costly physical copy, the development of a DT represented an economically viable alternative. This concept was then adopted by the United States Air Force for the maintenance of its fighter jets. With the emergence of Industry 4.0 and advances in hardware and software technologies, the DT has become a key element of digitalized industrial processes.

In terms of definition, those proposed respectively by Grieves and Vickers (2017) and by Tao (2018) often stand out. These two definitions, almost similar, identify the DT as being composed of three elements: the physical model, the digital model (DM) and the bidirectional communication link. Their work meets with broad consensus around the concept of the DT and also reveals that there is no universally accepted definition. Notably, the bidirectional nature of the communication between the physical and digital worlds is often interpreted in a unilateral manner.

Furthermore, Kritzinger et al. (2018) introduced three distinct notions: the DM, the digital shadow (DS) and the DT. Thus, the DM is a virtual representation without any direct communication link with its physical equivalent. The DS reflects the state and actions of the physical model, but the communication is unidirectional, from the physical to the digital. Finally, the DT is characterized by active bidirectional communication, enabling interactions between the digital and physical worlds. In what follows, two general notions will be addressed in order to better understand the DT.

7.1.1. *Maturity levels*

Yong-Woon et al. (2021) proposed a classification of DTs into five maturity levels, based on their characteristics and capabilities. This classification makes it possible to understand the evolution and potential of DTs in different industrial contexts.

Level 1, Mirroring Digital Twin, represents the most basic form of the DT. It is limited to a 2D or 3D graphical representation of the physical model, without any interconnection or behavior laws. It is essentially a static mock-up comparable to traditional CAD models. In continuity, Level 2, Monitoring and Control Digital Twin, establishes a rudimentary connection between the physical model and the DM. This connection is dedicated primarily to the supervision of the physical model. The collected data is used for basic logical analyses, such as process stoppage detection. This level corresponds to what Kritzinger et al. (2018) define as a "digital shadow".

Level 3, Modeling and Simulation Digital Twin, fully conforms to the definition of a DM connected in bidirectional mode to the physical model. This connection enables real-time synchronization and advanced simulation based on behavior laws. Although information is communicated instantaneously to the physical system, critical decisions generally remain the responsibility of human operators, for safety reasons.

Level 4, Federated Digital Twins, evolves beyond individual entities to form complex assemblies. In a future where each component will be accompanied by its own DT, it will be possible to build a global DT by assembling these multiple entities. This would enable a large-scale representation of complex entities such as cities. Level 5, Autonomous Digital Twins, will see DTs become fully autonomous. To this end, perfect synchronization between real and virtual elements is necessary. In this scenario, the DT would be capable of analyzing, making decisions and acting autonomously, reflecting in-depth interpretation and instantaneous responsiveness to events and needs of the physical system.

7.1.2. *Categorization*

Analysis of various DT examples from the literature reveals a diversity of concepts. Feedback in terms of design and deployment highlights the obstacles to overcome and integration strategies. These approaches can be synthesized into three categories.

The first category includes DMs, often presented incorrectly as DTs. They are mainly used for simulation, as they do not fulfill all the criteria that characterize DTs, particularly the absence of bidirectional communication between the DM and its physical equivalent. This underscores the importance of suggesting a universal definition that would make it possible to clearly distinguish DTs from DMs.

The second category is closer to the principles of DTs, notably through the integration, on the one hand, of a complementary Internet of Things (IoT) system and, on the contrary, through the establishment of bidirectional communication between the DM and physical systems. However, this configuration does not allow access to all advanced functionalities, such as dynamic 3D representation or extended and shared supervision. This is generally attributed to the fact that certain machines or industrial equipment were not designed to be integrated into a digitalized production environment. Furthermore, in most cases, their control system is closed and does not allow data feedback, hence the need to integrate a complementary IoT system.

The third category concerns machines or cyber-physical systems specifically designed to operate in a digitalized environment. These devices can thus be fully integrated into a global DT, facilitating the installation of bidirectional communications. Many machine tool and industrial equipment manufacturers now offer virtual models with certain characteristics to facilitate their integration into the DT of the entire production system. This evolution reflects a growing awareness among manufacturers of the need to make their equipment compatible with digital environments and to offer access to machine data. To this end, they adopt standardized protocols and 3D models compatible with the majority of digital platforms dedicated to the creation of bidirectional DTs.

Despite these advances, problems of interoperability, communication protocols and software persist, often requiring complementary work depending on the targeted functionalities. Numerous studies show that a majority of DTs are limited to a few functionalities without fully exploiting the potential of this technology, particularly in terms of intelligent supervision, operational flexibility, energy optimization, failure prediction or immersive training. The main reasons for this limitation are diverse. Firstly, there is technical complexity, due to the diversity of equipment and the requirements of certain functionalities which require access to real-time data and advanced calculations. Secondly, there is an economic dimension: a DT integrating all functionalities would be costly and would require regular technical support; hence, the importance of finding a balance between needs in terms of functionality and the expected performance also expressed on an economic scale.

7.1.3. *Operating modes*

The DT can operate in coupled or decoupled mode relative to the physical model. This distinction is essential for adapting the use of the DT to specific needs, particularly in terms of monitoring, analysis and control in various industrial scenarios.

In the coupled mode, the DT and the physical system are connected in real time or at very short intervals. This connection enables dynamic interaction between the two: data from the real system continuously feeds the DT, which can then predict and optimize the performance of the physical system in real time. Adjustments or actions generated by the DT are immediately sent back and applied to the physical system. This mode is particularly useful for dynamic supervision, performance monitoring and operations optimization.

Decoupled mode, on the contrary, means that the DT operates independently, often offline. It uses historical data to perform analyses, simulations or tests without directly impacting the real system. This mode is ideal for planning, training, new process development or scenario simulation before their implementation in the real world.

The choice between these modes depends largely on specific objectives, the criticality of operations, data processing capabilities and the need for real-time responses.

For example, a company may use coupled mode for real-time process control and switch to decoupled mode for development or improvement activities that do not require instantaneous feedback. Defining the operating mode from the design phase of DTs is crucial to align the approach with operational requirements, thus ensuring optimal integration and utilization.

In conclusion, these studies illustrate both the progress made and the persistent challenges in the field of DTs, highlighting the need for standardization of methodologies and an expansion of functionalities to better meet the evolving requirements of modern industry.

The next section introduces a generic design approach dedicated to DTs. It was implemented and experimented with at the ENSAM Industry 4.0 platform.

7.2. Design approach

Despite the numerous advantages and applications of the DT highlighted by recent academic and industrial work, its effective deployment requires a structured approach.

Thus, this work summarizes an innovative strategy broken down into three phases for the design and deployment of a DT, as shown in Figure 7.1.

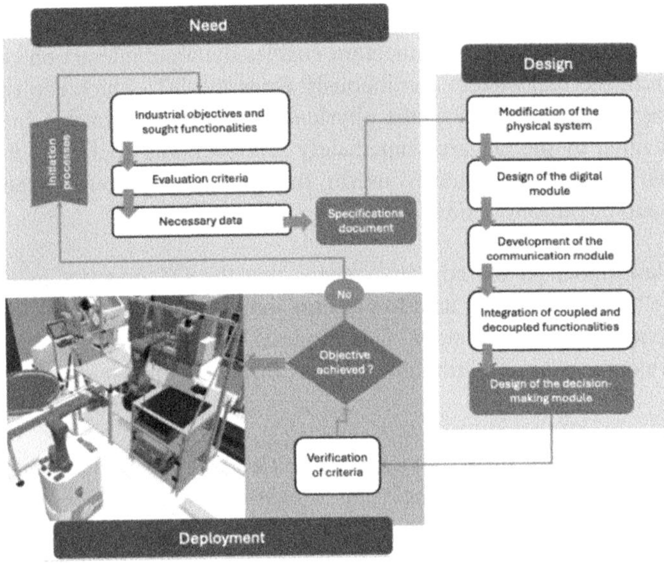

Figure 7.1. *Digital twin design process. For a color version of this figure, see www.iste.co.uk/benfriha/industrialprocesses.zip*

The first phase consists of defining the expected needs and objectives. It results in the creation of a specifications document defining the required functionalities, the necessary data and the validation criteria for the DT. The second phase, called the "design" phase, relies on five key sub-steps:

– modification of the physical system to be compatible with the DT requirements;

– design of a digital module, including a Multimodal Adaptive Interface (MAI) enabling interaction between operators;

– development of a communication module to facilitate data exchange with the physical system;

– implementation of the selected functionalities in coupled and decoupled modes;

– development of a decision module to support process supervision.

The third phase, the deployment phase, evaluates and quantifies the industrial performance improvements achieved through the implementation of the DT. If the expected performance is not achieved, the methodology suggests a return to the first phase, integrating the lessons learned from previous results and so on until satisfactory results are obtained.

7.3. Needs identification phase

As mentioned previously, the DT can integrate a range of functionalities, from the most basic and easily integrable to the most advanced, such as predictive maintenance, supervision, simulation, control, training, traceability, optimization, visualization, portability and security.

It is important to note that currently, no DT combines all these functionalities simultaneously. Companies must therefore prioritize the functionalities to be implemented. This section aims to formalize a specifications document specifying the targeted functionalities, the evaluation criteria, as well as the data necessary for their implementation.

7.3.1. *Identification of industrial objectives*

The development of a DT begins by defining the objectives, expressed in terms of functionalities and their execution mode, whether coupled, decoupled or both simultaneously. Furthermore, it is crucial to limit these functionalities to avoid making the DT excessively complex and difficult to deploy.

7.3.2. *Identification of evaluation criteria*

The evaluation of the DT requires the establishment of specific criteria to measure its adequacy to the defined needs. These criteria are essential during the final verification phase, confirming whether the DT meets expectations. Some are quantifiable, such as latency, acquisition frequency, number of frames per second (fps) and accuracy between simulation and reality, as well as the improvement of performance indicators relative to an initial baseline. Other functionalities are evaluated qualitatively due to the absence of easily measurable criteria. Finally, certain functionalities have their own evaluation criteria to measure their performance, such as optimization, which can be evaluated by the minimization of objective functions.

Functionalities	Criteria	Criteria values
Predictive	Failure detection rate and failure prediction time	Percentage and time
Security	Confirmed/avoided incidents	Actual number
Traceability	Traceability rate	Percentage
Scheduling	Objective functions: time, cost and energy	Minimization of functions: time, cost and energy

Functionalities	Criteria	Criteria values
Optimization	Improvement percentages	OEE, availability rate and workshop load rate
Supervision	Operations execution time and observed deviations	Time, stoppages, delays and slowdowns
Simulation	Accuracy and consistency	Error between simulation and physical <5%
Training	Proposed pathways	Skill acquisition time per pathway
Portability	Number of functionalities covered by portability	Data refresh rate, latency and availability
Visualization	Fidelity, latency and fps	Maximum time value, 30 fps minimum and geometric deviations

Table 7.1. *Simplified table of functionalities and associated criteria*

Table 7.1 summarizes some evaluation criteria for various DT functionalities. For each functionality, the evaluation criteria and associated units of measurement are suggested, thus providing a reference framework for evaluating the effectiveness of the DT during the deployment phase. This table is not exhaustive; other criteria may complement it.

7.3.3. *Identification of necessary data*

This step aims to identify the data necessary for each DT functionality, taking into account their operating modes. Data from coupled mode are those acquired in real time directly from cyber-physical systems and the Industrial Internet of Things (IIoT) system. They are transmitted to the DT, either directly or via the Manufacturing Execution System (MES), for immediate analysis and use. For example, machine position, their operational status or their energy consumption is typical data for this mode. In contrast, data from decoupled mode come from the data lake or previously performed simulations. These data are enriched over time thanks to data from coupled mode, including both archived data, machining programs and quality requirements. The compilation of these functionalities, evaluation criteria and data results in the proposal of a specifications document, presented in section 7.3.4.

However, it is important to understand that the choice of data flows between the DT and the MES will depend on the communication architecture and the functional structure of the production system.

In some situations, the DT will be prioritized for data collection and processing, while in others, the MES will handle this task, before transmitting the obtained information to the DT. This strategic choice is adopted by the company according to the roles and attributes it wishes to assign to the MES and the DT.

7.3.4. *Writing of the specifications document*

Table 7.2 presents a simplified example of a specifications document for a DT, including a few functionalities, as well as the evaluation criteria and necessary data. This document serves as a reference for the design and integration of the DT specific to industrial processes.

It is important to note that the specifications document is an evolving document, subject to modifications and adjustments according to desired evolutions or constraints encountered during the development of the DT.

Functionalities	Evaluation criteria	Required data
Predictive	Failure detection rate and failure prediction time	Multiphysical measurements on machines and post-failure reports
Security	Confirmed/avoided incidents	Videos, real-time positions of various moving components
Traceability	Traceability rate	Temporal and spatial recording of movements and transformations of all parts
Scheduling	Objective functions: time, cost and energy	Workshop load plan, production plan for new orders and prioritized functions
Optimization	Improvement percentages	Data on operations execution, machine breakdowns, setup time, incident reports, etc.
Supervision	Operations execution time and observed deviations	IoT data, progress of parts in production and communication network activity
Simulation	Accuracy and consistency	Production plan, geometric models of new parts and scheduling

Training	Proposed pathways	Learner profiles, production plan mock-ups, virtualized PLC and simulated failures
Portability	Number of functionalities covered by portability	All data recorded in the data lake, cyber-physical systems, sensors, PLCs, etc.
Visualization	Fidelity, latency and fps	Recording of positions of all moving components, indicators and machine status lights (red, white, green)

Table 7.2. *Simplified specifications document*

7.4. DT design phase

The design of the DT draws on the specifications document established previously and is inspired by the work of Kritzinger et al. (2018). This process is divided into five key steps: the modification of the physical system, the design of the digital module, the development of the communication module, the integration of functionalities in coupled and decoupled modes and the addition of a component for decision-making.

In order to make this approach pragmatic and compatible with the majority of existing industrial systems, the target system is assumed to rely on a hybrid pyramid architecture, described in Chapter 4. This hierarchical structure comprises five levels: at the base are machines and sensors, followed by Programmable Logic Controllers (PLCs), the Supervisory Control and Data Acquisition (SCADA) system, then the MES and finally at the top, the Enterprise Resource Planning (ERP) system.

7.4.1. *Modification of the physical system*

In existing infrastructures, access to certain data is limited, often due to the absence of adequate sensors or restrictions imposed by manufacturers on access to internal machine data. In order to ensure that a minimum of necessary data is available for DT functionalities, it is necessary to adapt the physical system. This may involve the installation of an IIoT system integrating sensors and a network interoperable with existing communication infrastructures.

This phase requires the implementation of a complete acquisition system, capable of capturing, analyzing and transmitting the necessary data. It may also require negotiations with manufacturers in order to access internal machine data. Furthermore, it is essential to design and implement a communication and storage architecture adapted to all collected data. Once this step is completed, the design of the DM and the communication module can begin.

7.4.2. *Design of the digital module*

The design of the DM is based on three components: the geometric model, the kinematic model and the MAI, enabling operators to interact with the DT. The geometric and kinematic models will constitute the immersive graphical space.

The geometric model provides a precise three-dimensional representation of the system, allowing visualization of its physical configuration in a virtual environment. The kinematic model complements the geometric model by adding movement and dynamism. In the absence of native models, these models must be recreated and assembled to form a faithful digital mock-up of the system. Finally, the MAI optimizes human–DT interaction by displaying various functional indicators. It allows simultaneous visualization and manipulation of the physical system and its DT in coupled mode or only the DT in decoupled mode.

It is essential that the DM faithfully reflects physical reality. 3D models are often based on initial plans that may evolve over time. Consequently, regular updates of the DM are necessary to guarantee accuracy with the physical system.

7.4.3. *Development of the communication module*

This section examines the main industrial architectures, such as service-oriented architecture (SOA), event-driven architecture (EDA) and computer-integrated manufacturing (CIM), in order to propose an optimized configuration for the DT. Each of these architectures presents specific advantages and responds to particular needs, directly influencing data valorization and availability. The proposed architecture, shown in Figure 7.2, combines the strengths of SOA, EDA and CIM while being designed to meet the specific requirements of the DT.

The communication module of the DT ensures the link between the real world and the virtual world. For an effective design, this module must integrate several interfaces: one to interact with the MES, another to manage physical sensors and a third to connect to the data lake.

Furthermore, the communication module plays an essential role in the portability of the DT. The choice of communication protocols and technologies conditions the possibility of remote access to data. For example, data collected by sensors can be made accessible externally by means of an API, generally using TCP/IP protocols over the Internet. When the DT is remote, it is crucial to guarantee reliable communication to update the data lake in real time, from which the DT retrieves necessary data or information. However, this portability often encounters strict corporate cybersecurity policies, which must manage the risks of data leaks and external threats.

The architecture shown in Figure 7.2 combines the flexibility of SOA and EDA architectures with a CIM pyramid structure, thus offering an innovative solution for dynamic interaction with the DT. The different modules of the DT operate as autonomous services within the SOA architecture. These services exchange information within the EDA architecture, where the generated data passes through the communication module to the appropriate computing module, in order to be analyzed and interpreted in the form of events or information. The events are then transmitted to the decision-making module, responsible for triggering necessary actions or not. As for the information, it is displayed simultaneously in the immersive graphical space of the DT and in the MAI. Furthermore, this interface allows operators and supervisors to interact, either only with the DT for simulation needs in decoupled mode or simultaneously with the DT and the physical model in coupled mode. Moreover, bidirectional portability offers access to certain attributes of the DM, either by means of a remote connection to the local virtual server or to the cloud.

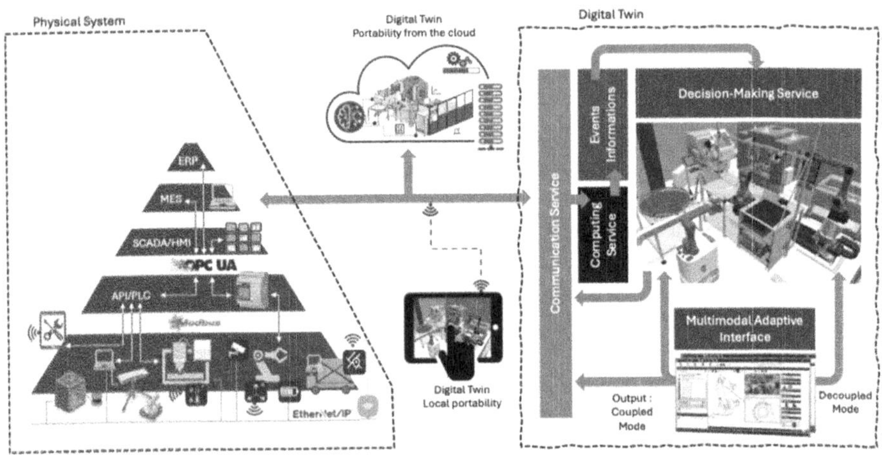

Figure 7.2. *Architecture dedicated to the digital twin. For a color version of this figure, see www.iste.co.uk/benfriha/industrialprocesses.zip*

In summary, the communication module is a key component in the DT architecture, facilitating the collection, processing and distribution of data or actions.

Its design requires rigor to ensure the overall efficiency of the DT and its ability to meet the specific needs of the production environment.

7.4.4. *Integration of functionalities*

This step consists of deploying the predefined functionalities, capable of operating in coupled mode, decoupled mode or simultaneously. The establishment of a general methodology for their integration remains complex because these functionalities are abstract and often interdependent. However, their effects manifest concretely through the behavior of industrial systems. Figure 7.2 shows an implementation logic based on a pipeline-type process. To this end, the functionalities extract data through the communication module, which is then processed by algorithms specific to each functionality and referenced by the computing service. These algorithms generate two types of output: information intended for the MAI, enabling interaction with the operator; and events transmitted to the decision-making module to control the physical system.

7.4.5. *Design of the decision-making module*

The design and implementation of decision-making modules capable of generating actions are essential to enable the DT to actively influence the physical system. To this end, the integration of artificial intelligence algorithms is recommended, especially if the objective is to confer autonomous supervision attributes to the DT for production systems. The choice of analytical tools depends on the complexity of the decision-making processes to be implemented. However, their use is not exclusive, and other conventional methods, such as decision trees, can also be considered.

Decision trees provide for the programming of responses associated with various events and their possible combinations. They are relatively simple to implement but require an in-depth analysis of potential scenarios and corresponding actions. However, they can prove limited for complex systems and may not react adequately to unforeseen scenarios. In contrast, neural networks constitute an advanced alternative. This approach requires a prior training phase with a wide range of data from the various functionalities. Although more complex and requiring a longer development time, it offers superior relevance and scalability, fitting into a long-term vision of the DT.

Once developed, the decision module transmits its directives to the physical system via the communication module and, in parallel, to the three-dimensional graphical space. This integration contributes to finalizing the design of a bidirectional DT capable of reacting according to data, events and production priorities.

7.4.6. *Synthesis*

This approach led to the creation of a functional DT, tested on the Industry 4.0 platform. Figure 7.3 presents an overview of this DT, which includes a video stream allowing live visualization of the physical platform, as well as an MAI offering bidirectional interaction between operators and the physical system in coupled mode. In decoupled mode, interaction is limited to the immersive graphical space, notably to meet simulation needs.

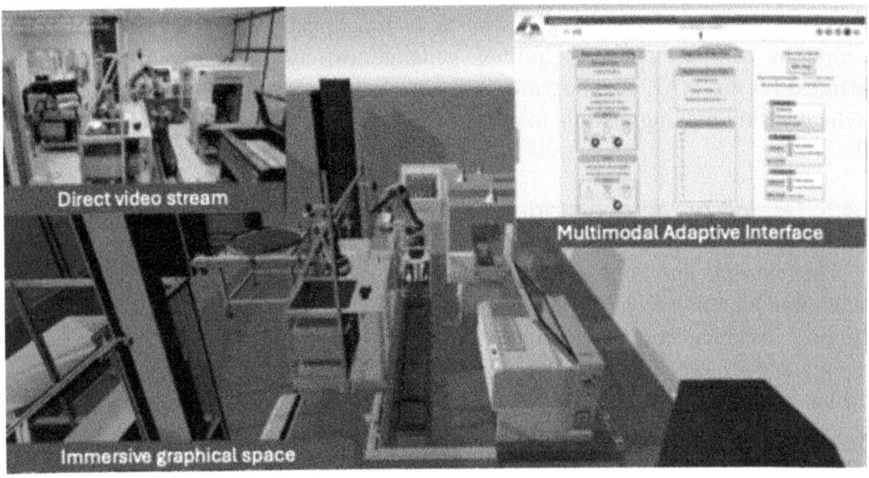

Figure 7.3. *Digital twin of the Industry 4.0 platform. For a color version of this figure, see www.iste.co.uk/benfriha/industrialprocesses.zip*

Depending on the context, the five steps of the approach can be modified, adapted or supplemented. As a reminder, the first step consists of adapting the physical system to make it compatible with the DT, which includes adding sensors and modifying the existing infrastructure to ensure the collection and transmission of necessary data. The second step concerns the design of the digital module, which includes a three-dimensional or immersive graphical space and an MAI, thus allowing the DT to interact effectively with the physical system.

Next comes the development of a communication module to establish a communication channel between the DT and the physical system. The analysis of various architectures, such as EDA, the CIM pyramid and SOA, reveals the absence of a standard architecture, thus highlighting the need to design a hybrid architecture. The proposed architecture combines the principles of the CIM pyramid architecture and SOA, making it possible to consider the DT as a set of modules offering specific services. This approach is reinforced by the EDA architecture, which effectively supports the decision-making process.

The fourth step consists of integrating the priority functionalities and their calculation algorithms, thus meeting the performance sought by the company. Finally, the fifth step concerns the control of the physical system, which relies on the decision-making module, information and generated events.

7.5. Deployment and verification phase

Before moving to the deployment phase, which begins with the development of the specifications document defining the DT to be designed, a brief presentation of the environment where the experiments are conducted will be introduced. Indeed, this approach is carried out on the Arts et Métiers Industry 4.0 platform, dedicated to research in the field of digitalization of industrial production systems. This technological platform, detailed in Chapter 5, will be briefly presented in section 7.5.1.

7.5.1. *Experimentation platform*

The Industry 4.0 platform is a research site dedicated to the digitalization of industrial production systems. It is distinguished by a full-scale production line, integrating various equipment, where scientific experiments are conducted on methodological, technological and algorithmic themes. The production line comprises three computer numerical control (CNC) manufacturing machines: a milling machine, a lathe and a laser cutter, arranged in an integrated configuration, thus making the manufacturing process more coherent while optimizing operations scheduling.

To complement this configuration, two six-axis industrial robots are integrated, as shown in Figure 7.4. One of these robots is equipped with a seventh translation axis, allowing it to interact with almost all the equipment on the platform. The second robot, fixed to the distribution table, has two main missions: to feed the production line with raw parts and to evacuate finished parts to the output. Its second mission, more complex, consists of collaborating with the first robot in a shared workspace, thus optimizing operational efficiency. Furthermore, a collaborative robot mounted on an autonomous mobile platform has been added to increase flexibility by allowing parts to be moved beyond the fixed limits of the production line.

In this context, a 3D optical scanner, placed near the robots, allows the MES to organize and activate geometric and dimensional conformity control of parts at different stages of the processes. To this end, the robots load the scanner with the parts to be inspected and remove them at the end of the operation to reintegrate them into the manufacturing flow. The inspection periods are adjusted according to the modeling of observed deviations. Thus, the MES can bring inspection periods closer if deviations are increasing or space them out if deviations are stabilized. This integration, both technological and computational, contributes to increasing process autonomy and, consequently, to reinforcing the digitalization of the production line.

Figure 7.4. *Platform 4.0. For a color version of this figure, see www.iste.co.uk/benfriha/industrialprocesses.zip*

Finally, the platform has evolved through several phases to achieve a high level of digitalization. Its hybrid and multi-layer architecture reflects a model widely adopted by companies, taking into account existing production infrastructures.

7.5.2. *Specifications document*

The objective is to establish a simplified specifications document, limited to three functionalities, which will serve as a basis for the DT design phase. Similarly, a single evaluation criterion has been retained for each functionality in order to facilitate the approach.

7.5.2.1. *Objectives and functionalities*

To avoid complicating the deployment approach, only three key functionalities have been selected. The first concerns the active security of digitalized production sites because the increased presence of fixed robots and mobile robots can generate critical situations involving humans. The second functionality concerns the supervision of production operations, notably through the dynamic detection of collisions between robots or with any other physical object. This functionality is essential because operations are supervised dynamically and in real time and any change can lead to collisions. Finally, the third functionality concerns visualization, expressed in terms of temporal accuracy between the physical system and that simulated by the DT.

7.5.2.2. *Definition of evaluation criteria*

This section establishes criteria to evaluate the effectiveness of the DT during its deployment. Each functionality, whether coupled or decoupled, is assigned specific criteria. Table 7.3 summarizes for each functionality the assigned criterion and its value.

Functionalities	Evaluation criteria	Values
Supervision	Minimum detection time	1 second
Security	Maximum acquisition rate	100 ms
Visualization	Maximum temporal error	5%

Table 7.3. *Summary table for the deployment phase*

7.5.2.3. *Identification of necessary data*

For each functionality, the types of necessary data are specifically identified. Concerning supervision, the DT requires information on dynamic operations scheduling as well as real-time actions coming from decision modules. Furthermore, collision detection between robots requires real-time access to information relating to the dynamic configuration of robots.

For the security of production lines and their environment, the DT has priority access to video streams from cameras covering all workspaces, particularly those equipped with moving components.

As for visualization, 3D models of equipment and their layout are required. However, these CAD models are not always designed to be animated and may lack kinematic links. It is therefore imperative to supplement these models to integrate the necessary functionalities and obtain a kinematic model whose behavior can be simulated.

7.5.3. *Design of the DT*

This section implements the five steps of the design approach presented previously. The first consists of adapting the Industry 4.0 platform to the requirements of the functionalities described in the specifications document. This step requires hardware and software adjustments to facilitate the integration of the DT. The second step concerns the development of the digital module, including the immersive graphical space, where the production line can be visualized virtually, as well as an MAI for control. Next comes the communication module, which ensures the link between the digital module and the physical system.

The following step consists of integrating the functionalities under different operating modes, ensuring that they are properly characterized. Finally, the decision-making module is developed. This module allows the DT to actively supervise the physical system thanks to the analysis of upstream data.

7.5.3.1. *Adaptation of the Industry 4.0 platform*

The upgrading of the Industry 4.0 platform focuses on two main aspects: the integration of an IoT network with new sensors, including cameras, and the implementation of an architecture adapted to the collection of necessary data. The choice of communication protocols depends on the sensors used and the existing IT infrastructure. Modern protocols such as MQTT and OPC UA are recommended to meet the needs of the selected functionalities. Furthermore, data from PLCs and the MES, already accessible via OPC UA, are available for exploitation.

7.5.3.2. *Design of the digital module*

The choice of development platform for the digital module of the DT is crucial for the realization of the project. In this regard, video game engines such as "Unity" or "Unreal Engine" offer a conducive and flexible environment for the development of DTs. These engines generally integrate a graphics engine for creating 3D environments and a physics engine to manage interactions and collisions. Furthermore, they allow complex behaviors and interactions to be programmed, which makes them particularly suitable for creating interactive and dynamic DTs.

For this experiment, the choice was made for "Unity". The latter offers a combination of advanced graphical capabilities, integrated physics management and the possibility of custom programming. Unity constitutes a robust platform for developing a DT, enabling immersive user interaction and a faithful representation of the physical system, as shown in Figure 7.5.

Figure 7.5. *Overview of the digital module development environment. For a color version of this figure, see www.iste.co.uk/benfriha/industrialprocesses.zip*

Without going into the details, the Unity environment is composed of six spaces:

– Space 1, Scene View, makes it possible to geometrically model the workshop and to visually explore the interior of the scene from different angles.

– Space 2, Game View, offers a preview of what the operator can see when interacting with this virtual environment.

– Space 3, Hierarchy, lists all objects present in the current scene and allows the properties of these objects to be defined, such as kinematic links, degrees of freedom and physics.

– Space 4, Inspector, displays the details and properties of the currently selected object, including materials, as well as its position, rotation and scale in 3D space.

– Space 5, Project, serves as an explorer for Unity project files and folders, such as scripts, textures, 3D models and colliders, which allow collisions between different objects to be detected.

– Space 6, Console, displays messages, warnings and errors.

In the Scene View, 3D geometric models of production lines are generally available and can be adapted for use in the DT. These models must accurately represent the current state of the production line, taking into account modifications and additions made since their creation. When the necessary 3D files are not available, they can be generated by reverse engineering or by 3D scanning, depending on the resources and precision required. Kinematic models make it possible to animate these geometric models by adding movement and dynamism. The creation of these models can be complex, especially if the existing geometric models do not have clear kinematic links between their components. In this case, it may be necessary to cut and reassemble the geometric models to obtain realistic animations.

Furthermore, the design of the adaptive and modular interface is based on the functionalities defined in the specifications document and user needs. It can integrate existing human–machine interface elements or be developed from scratch. The display of information can vary dynamically depending on the position and orientation of the operator, thus creating a multi-scale user experience. For the Industry 4.0 platform, this interface is designed to adapt to different types of screens and devices, including immersive technologies such as virtual reality (VR). This allows stakeholders to interact immersively within the production environment.

Finally, interaction management within Unity is carried out using C# scripts. However, Unity does not have its own integrated development environment (IDE) for C#. For this project, Visual Studio is used, offering a robust and complete programming environment, with advanced tools for code editing, debugging and fluid integration with Unity.

7.5.3.3. *Development of the communication module*

The development of the communication module is based on the OPC UA interface, used to interact with the supervision system (MES) and the PLC of the Industry 4.0 platform. OPC UA facilitates real-time data exchange between the MES, acting as a server, and the DT, operating as a client. This protocol ensures fluid and fast transmission of information such as the production plan, electrical data, and robot positions. In contrast, G-codes do not require a specific interface, as they are directly accessible from the hard drives of the CNC machine control units.

However, data acquisition from the physical system can cause significant latency due to the passage of this information through several intermediaries, notably the PLC and the MES. This latency constitutes a major obstacle, particularly for applications requiring real-time responses, such as precise tracking of robot positions. To circumvent this problem, direct communication between the equipment and the digital module has been established. The chosen approach uses Windows Communication Foundation (WCF), where each robot is equipped with its own WCF server, while the digital module operates with clients that connect to these servers. As for G-codes, the programming files of CNC machines are directly accessible by the digital module as soon as their location on the network is identified.

Furthermore, several difficulties may arise during experimental phases. A drastic drop in the fps rate has notably been observed. This phenomenon is caused by the synchronous execution of Unity scripts, where each line of code waits for the previous one to finish executing before executing. Similarly, when the script must establish communication with equipment, execution is temporarily suspended, causing a delay of several tens of milliseconds due to latencies related to these exchanges. Given the large number of interactions between Unity and the physical system, this situation considerably reduces the fps rate of the DT, dropping from 45 to 12. To solve this problem, it is crucial to implement asynchronous communications for all interfaces, in order to avoid these interruptions and maintain optimal application performance.

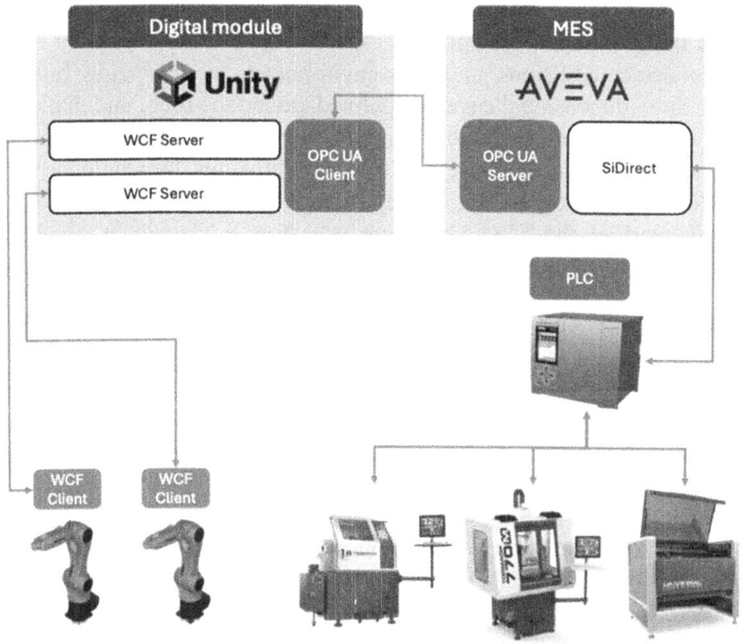

Figure 7.6. *Communication architecture of the Industry 4.0 platform digital twin. For a color version of this figure, see www.iste.co.uk/benfriha/industrialprocesses.zip*

In conclusion, the communication module constitutes a crucial interface, enabling the integration and exchange of data between the physical system and the DT.

Figure 7.6 presents the communication architecture adopted for the Industry 4.0 platform. The next stage of the project aims to develop a multimodal interactive interface that will improve dialogue with the DT, enriching the user experience and offering an in-depth real-time overview of the production process.

7.5.3.4. *Integration of functionalities*

The first functionality to integrate allows real-time collision detection to be apprehended. Indeed, operations orchestrated in dynamic mode can generate critical situations, particularly in terms of physical collision. The digitalization of industrial production systems implies a sharing of the workspace between different physical components in motion. Thus, the MES can, depending on unforeseen events or changes in production priorities, modify the scheduling of operations or add new ones if necessary.

The objective is to anticipate any potential collision between the robot and other elements of the production system. The capabilities of Unity's physics engine are exploited to detect collisions in a 3D environment, by analyzing interferences between the three-dimensional envelope defined around the robot and that of another component of the system. The algorithm projects this envelope with a lead time on actual production in order to prevent any critical situation. Depending on the time before collision detection in 3D space, corrective measures can be envisaged, including an alert in the MAI, an immediate stop, a slowdown or a temporal rescheduling of operations involving potential collisions.

Robots, industrial equipment or any other moving component is surrounded by an envelope consisting of a 3D point matrix, defined with a precision of 0.1 mm. This precision is realistic in many industrial applications. Advanced algorithms analyze this data to detect collisions, by calculating whether objects share close or identical coordinates. This sophisticated method offers real-time prediction to prevent any impact that could generate critical situations.

The second integrated functionality concerns the security of operators evolving in a robotized production site. The objective is to endow the DT with a capacity to detect and characterize any human intrusion into the workspace of machines and robots, in order to make decisions in accordance with the observed situation. The choice to prioritize this functionality is justified by the necessity of having active security of integrated production systems.

To this end, the algorithm embedded in the "computing service" of the DT analyzes in real time the images captured by cameras positioned at height, which can provide overall views of the production line. It also explores information related to the various production states, whether it be stopped, in progress, under maintenance, etc. The general architecture that allows this functionality to be modeled is represented in Figure 7.7. In terms of computational and algorithmic resources, this application can operate with relatively simple tools. A basic decision tree, coupled with a human detection module available in several open-access libraries, should suffice to ensure its effectiveness.

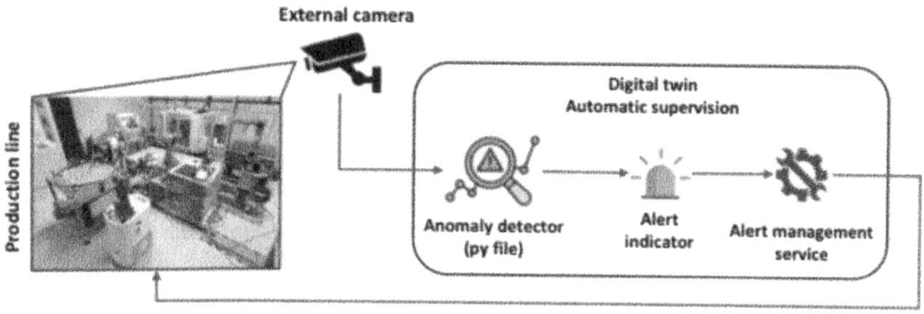

Figure 7.7. *Overall architecture of the supervision functionality. For a color version of this figure, see www.iste.co.uk/benfriha/industrialprocesses.zip*

Figure 7.8 shows the integration of this functionality within the DT of the Industry 4.0 platform. The upper part of the figure presents a normal situation without intrusion; a green light thus signals the absence of human presence near the machines. In contrast, in the lower part, a human presence is detected by the camera, and the DT indicates this by a red light visible in the immersive graphical space.

The DT, being bidirectional, not only allows the interpretation of upstream data relating to anomalies and alerts but also the sending of control signals directly to robotic systems, without passing through the MES. This approach proves particularly effective in critical situations, where every millisecond counts, while highlighting the complexity of communication network interoperability.

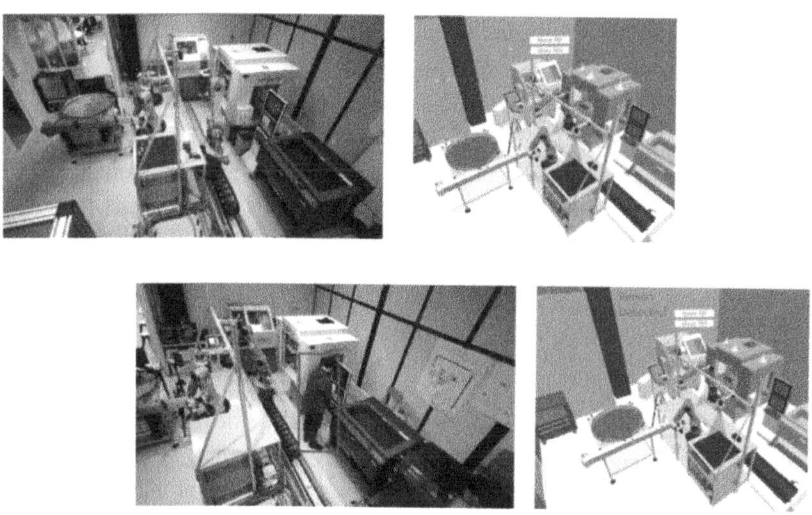

Figure 7.8. *Example of person detection. For a color version of this figure, see www.iste.co.uk/benfriha/industrialprocesses.zip*

The third functionality concerns the ability to visualize the production system in different modes. This functionality relies, on the one hand, on the accuracy and attributes of the 3D models, which produce behavior faithful to the physical model, and on the other hand, on the relevance of data generated by production as well as the availability of information relating to the manufacturing process. Once the DT is established, visualization tools such as interactive dashboards, augmented reality (AR) or VR offer users a multi-scale immersion in the production system. For example, AR visualization can superimpose digital information onto a live view of the physical system, allowing operators to observe performance in real time and quickly detect anomalies or points of failure.

Within the framework of the Industry 4.0 platform, the supervision functionality updates the immersive graphical space and the multimodal interactive interface in order to faithfully represent the state of the real system. Robot positions are transmitted in real time thanks to WCF communications, while information on the state of machine tools, although more complex to collect, is represented by simple animations. Similarly, door sensors provide instantaneous data on the open or closed state of machine doors.

For the localization of parts in the manufacturing process, the DT relies on the MES, which manages real-time operations scheduling on robots that transfer and store parts at different phases of the process, and on cameras, which provide complementary information. The convergence of this information towards the DT gives it the ability to faithfully and in real time trace the spatiotemporal positions and manufacturing stages of all parts that travel through the production line. Furthermore, this allows certain aspects, such as supervision, monitoring and optimization, to be organized in shared and remote mode.

7.5.3.5. *Decision-making module*

The decision-making module confers upon the DT operational autonomy. Within the framework of this experiment, and to simplify the design of this module, an approach based on predefined scenarios, known as a "decision tree", was favored, rather than artificial intelligence methods which would have made it possible to increase autonomy and reinforce the evolutionary aspect of the decision-making process.

This approach consists of anticipating all possible combinations of events and programming specific responses for each scenario. It involves an in-depth analysis of the different events likely to occur, requiring immediate intervention. This method requires exhaustive knowledge of processes and the production system in order to define the appropriate responses for each combination of events.

For this experiment, a decision tree was developed based on a potential simultaneous failure of all machines, leading to a total production interruption. Other events are handled independently according to their level of criticality. Thus, events requiring immediate intervention are treated first according to their order of priority. The decision tree offers a clear view of how the DT manages different scenarios, making decisions based on collected events. This approach, simple to deploy, allows the DT to activate appropriate decisions in response to known situations. Figure 7.9 shows a basic example of a decision tree designed to manage a potential collision involving a robot.

There are, however, sophisticated methods based on artificial intelligence. These methods, which were not experimented with here, attribute to the DT a superior capacity for analyzing events in dynamic mode, allowing it to detect those that generate undesirable events and make appropriate decisions according to the criticality of events. Nevertheless, their development and deployment are time-consuming, requiring specific skills.

The next step consists of evaluating the DT and verifying its performance according to the three functionalities previously identified: security, supervision and visualization. The objective is to test it under varied conditions, observe its interactions with the physical system and evaluate its capacity to make decisions.

7.5.4. *Validation*

Regarding the security functionality, the tests carried out reveal a significant time difference in terms of DT responsiveness after the detection of human presence near a machine. Indeed, in the case of sending a stop signal to the robot via the MES, the response time is 92.54 ms. In contrast, in the case of direct sending of a stop signal to the robot, the response time is 10.23 ms. This difference in reaction time is attributable to the diversity of communication protocols used between the DT, the MES and the PLC. These protocols differ in their transmission characteristics. In particular, the MES, which controls the rest of production, may not prioritize the transmission of the alert signal if the robot is already engaged in an operation.

Concerning supervision, the experiment aimed at validating collision detection by the DT consists of placing an obstacle on the trajectory of a robot. This obstacle is also integrated into the virtual environment of the DT, positioned exactly at the same location as its physical equivalent. During the experiment, a movement is commanded to the robot, from the DT, so that its planned trajectory intersects the position of the obstacle. The experiment was conducted by varying the robot speed between 10% and 100% of its maximum speed, which is 2 m/s, with a prediction time set to one second.

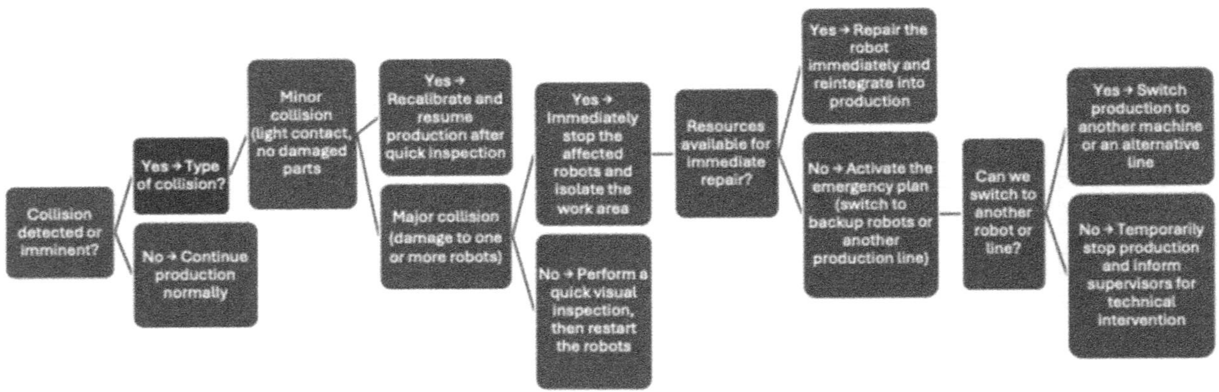

Figure 7.9. *Example of decision tree for the decision-making module*

For each speed, the experiment is repeated five times to guarantee its reliability. The results show that at speeds of 10% and 50% of the robot's maximum speed, the DT succeeds in detecting the obstacle and stopping the robot before it collides. In contrast, when the robot speed is above 50% of its maximum speed, the DT cannot interrupt the movement in time, thus resulting in a collision. This information can be exploited by the supervision system to reparameterize robot speeds when they operate in a shared workspace, where the risk of collisions is significant.

As for the visualization functionality, the experiment was limited to verifying the temporal accuracy between real trajectories and those simulated by virtual machines. Table 7.4 compiles the deviations measured between real and simulated trajectories for different missions, such as returning to the parking position or transferring a part to the temporary storage space. The measured deviations remain below the 5% threshold defined in the specifications document.

Trajectories	Real time	Simulation time
RF_Parking	10.69 s	10.2 s
RF_PosTransStore	17.94 s	17.34 s
RM_Parking	8.33 s	8.13 s
RM_PosTransStore	13.68 s	13.52 s

Table 7.4. *Comparison between real and simulated trajectory times*

In conclusion, this work has demonstrated the value of a structured approach for the design of DTs. The experiments highlighted the need to define upstream the needs and expected attributes. Indeed, the DT can occupy different functions, ranging from a simple visualization tool to an intelligent autonomous supervision system. The experiments conducted on the Industry 4.0 platform demonstrate that the expected performance objectives of the DT are achievable. However, they also highlight the technological challenges and the need for multidisciplinary skills.

8

Clouds for Industry

In Industry 4.0, the cloud plays a central role in the digital transformation of industrial processes. Three main typologies of cloud usage stand out in this context: cloud manufacturing, cloud outsourcing and cloud computing.

Cloud manufacturing enables optimized and flexible management of manufacturing processes by outsourcing certain production resources to the cloud. This allows companies to share, collaborate and coordinate their activities across distributed platforms. Cloud outsourcing, for its part, focuses on the outsourcing of certain services or infrastructures related to production or industrial management, such as inventory management or IT infrastructure. Finally, cloud computing, often used for computational analysis of industrial data, enables the processing of large quantities of data thanks to intensive computing capabilities, essential for tasks such as numerical simulation or process optimization.

In addition to these three main typologies, there are other forms of cloud which, although distinct, can be considered as subcategories or extensions of these models. For example, cloud storage is an essential element for cloud computing, as it allows large quantities of data to be stored before they are analyzed. Cloud storage plays a crucial role in industrial data management, particularly for connected sensors that generate large quantities of data, such as vibratory or acoustic signals.

The Hybrid Cloud fits within the logic of cloud outsourcing by enabling partial outsourcing. This model combines public and private cloud infrastructures, thus offering a flexible solution for companies that wish to optimize their resources while keeping certain critical data on a private cloud for security reasons.

The Edge Cloud (or edge computing) is particularly relevant within the framework of cloud manufacturing, as it brings computing capabilities closer to the industrial devices themselves. This improves responsiveness and real-time management of operations, thus reducing latency. Furthermore, in environments where delays are critical, such as real-time detection of potential collisions, edge cloud proves to be an essential extension of traditional cloud services.

Finally, the IIoT cloud (Cloud for the Industrial Internet of Things) can be considered as a specific application of cloud manufacturing and cloud computing. It enables the management of connected devices within an industrial infrastructure, thus facilitating the collection and analysis of data from sensors and connected machines, thereby optimizing production.

In summary, the three main typologies of the cloud – cloud manufacturing, cloud outsourcing and cloud computing – represent the pillars of digitalization of industrial processes for a digitalized industrial process. The other forms of cloud, such as cloud storage, hybrid cloud, edge cloud and IIoT cloud, integrate into these typologies as subcategories, offering specific solutions for varied needs, notably data management, infrastructure flexibility or industrial system responsiveness.

8.1. The main clouds for industry

This section provides more details on each of these three main typologies, addressing their applications, their advantages, as well as their impact on manufacturing methods, resource optimization and large-scale data analysis.

8.1.1. *Cloud manufacturing*

Cloud manufacturing represents a profound transformation in the way companies manage and organize their production processes. It offers a natural solution for moving from production-centric manufacturing to service-oriented manufacturing. This model is based on flexible access to distributed production resources. By using a cloud platform, companies can access manufacturing capabilities located in different places and exploit them according to their specific needs.

Typical applications include on-demand production, where manufacturing capabilities are allocated in real time according to market fluctuations or order personalization. This makes it possible to adopt more agile production models, such as just-in-time, which reduce unnecessary inventory and associated costs.

The advantages of cloud manufacturing are numerous: increased flexibility, reduced infrastructure costs and access to advanced technologies without requiring heavy investments in equipment or specialized personnel. Furthermore, the cloud fosters collaboration between different industry players, facilitating co-manufacturing or collaborative production between multiple sites or partners. Users can request services throughout the product lifecycle, from design to manufacturing.

The cloud manufacturing platform also offers search, intelligent mapping and recommendation services while providing the functionalities necessary for manufacturing process management.

In terms of impact on manufacturing methods, cloud manufacturing facilitates process integration, enabling smart factories to better manage production phases, adapt quickly to changes in demand and optimize machine utilization. This results in improved overall productivity and reduced downtime.

In conclusion, cloud manufacturing constitutes a distributed system, composed of an interconnected set of virtualized manufacturing resources and capabilities. It integrates intelligent management capabilities and offers on-demand services. This model supports the entire value chain by meeting the varied needs of users throughout manufacturing processes, whether for material or immaterial resources.

8.1.2. *Cloud outsourcing*

Cloud outsourcing allows companies to outsource their IT infrastructure by entrusting the management of their computing needs to specialized providers. This includes server hosting, the use of Infrastructure as a Service (IaaS) to rent complementary hardware resources, networks or even application development platforms through Platform as a Service (PaaS). These solutions offer flexibility and scalability, allowing companies to adjust their resources according to needs without investing in expensive hardware.

Companies can also outsource data storage for large-scale backup and entrust IT security management to specialized providers offering advanced solutions such as intrusion detection and protection against cyberattacks. Furthermore, software such as Enterprise Resource Planning (ERP), Computerized Maintenance Management System (CMMS) or Customer Relationship Management (CRM) can be outsourced through the Software as a Service (SaaS) model, thus offering access to updated, shared and distributed tools.

This approach frees companies from the constraints related to physical infrastructure management while offering them flexible and scalable IT resources.

According to Grand View Research (2023), the cloud outsourcing market was estimated at $90.28 billion USD in 2022, with an expected annual growth of 14.2% until 2030, driven by the expansion of IaaS, SaaS and PaaS services, as shown in Figure 8.1.

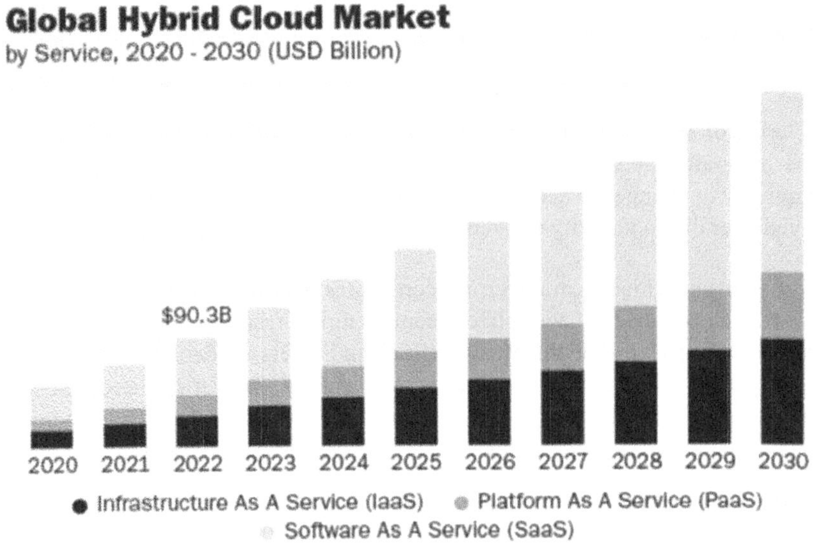

Figure 8.1. *Study conducted by Grand View Research. For a color version of this figure, see www.iste.co.uk/benfriha/industrialprocesses.zip*

The advantages are numerous: reduction of maintenance costs and initial investments, increased flexibility and the ability to evolve or adapt almost instantaneously according to needs. The cloud also offers high security standards, reassuring companies that must protect critical data. By outsourcing certain functions, companies optimize their resources and can focus on their core business while accessing the latest technologies such as artificial intelligence (AI) and machine learning (ML), without requiring heavy hardware or software updates.

In conclusion, cloud outsourcing represents a major strategic lever for more efficient and flexible resource management in Industry 4.0. By outsourcing their IT services, companies focus on innovation and customer satisfaction while benefiting from the economic and technological advantages offered by the cloud. It indeed allows access to expensive technologies at lower cost, thanks to a usage-based business model.

8.1.3. *Cloud computing*

One of the main advantages of cloud computing for industry is access to high-performance computing (HPC) resources on demand. Sectors such as automotive, aerospace and pharmaceuticals often require significant computing power for complex simulations and analysis of large volumes of data.

In Industry 4.0, where factories generate large quantities of data, the cloud offers computing capacity that is both sustained and economically accessible. Needs in terms of modeling, numerical simulation, predictive maintenance and supply chain optimization rely on these resources. For example, data from IoT sensors can be exploited to anticipate any performance degradation, thus limiting interruptions or production losses. Similarly, the cloud also provides access to advanced technologies such as AI and ML. These platforms integrate ready-to-use AI/ML services, accessible even without specific expertise.

In conclusion, cloud computing offers unparalleled flexibility and computing power, fostering significant advances in areas such as HPC, data analysis, application development and AI. Industrial companies can thus optimize their resources and improve their performance. However, this transition requires rigorous management of outsourced data to guarantee its security and confidentiality.

8.1.4. *Methodologies and frameworks*

Unlike the concepts of cloud manufacturing and cloud outsourcing, which have been widely explored in the literature and by technology solution providers, cloud computing has not generated abundant scientific and methodological production. This is explained by the specificity of the needs that this technology must satisfy, which includes both a business requirement, often centered on improving operational performance, and the nature of data generated at each stage of the process. In this context, two frameworks considered relevant, those of Wang et al. (2017) and Terrazas et al. (2019), will be presented below.

The framework developed by Terrazas et al. focuses on the adoption of cloud computing in manufacturing processes. It consists of two modules: Mc-Monitor and Mc-Analytics. The Mc-Monitor module is an integrated data management solution, designed to continuously collect, preprocess, transmit and store machining data. In contrast, the Mc-Analytics module constitutes an analysis framework based on cloud computing technologies, enabling the extraction of actionable information from this data. Figure 8.2 shows the physical, network and cybernetic components included in the Mc-Monitor and Mc-Analytics modules.

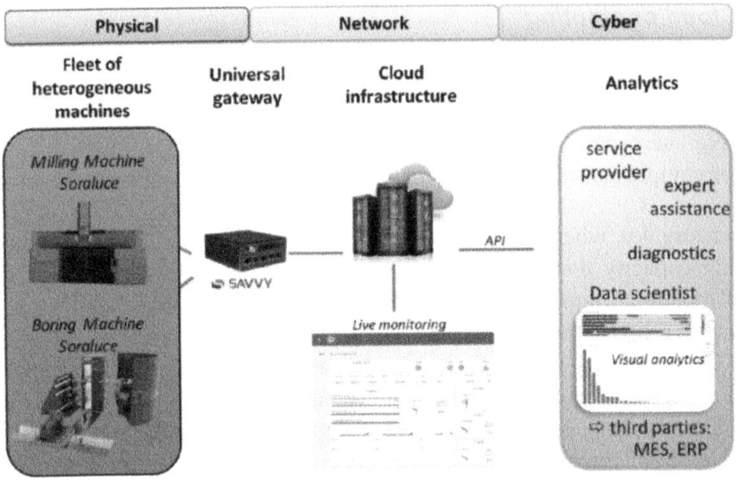

Figure 8.2. *The framework proposed by Terrazas et al. (2019)*

To interpret data from heterogeneous machines, a universal gateway called the Savvy Smart Box is used. In this system, data from sensors is retrieved, formatted and sent to a cloud platform called Savvy Industrial Cloud, using Machine-to-Cloud protocols. This data is then accessible via a REST API (Representational State Transfer). Mc-Analytics exploits this data to provide end users with specialized assistance and in-depth analyses.

The data flow in Mc-Monitor follows a specific process: Data → Gateway (Savvy) → Cloud Platform (Savvy Industrial Cloud) → Apache Storm on the cloud for preprocessing. In Mc-Monitor, data is classified into online and offline types. For offline data management, a document database is used to store structured and unstructured data. Regarding online data (measurements), a message queuing technology is implemented, as it guarantees message publication and consumption.

Mc-Analytics is structured into several layers, as shown in Figure 8.3. The first, the knowledge layer, centralizes and retrieves manufacturing data, whether from online or offline sources, through the Mc-Monitor module. The second, the analytical layer, aims to systematically extract knowledge using scripts to analyze indicators such as energy consumption. The third, the application layer, orchestrates analytical processes by controlling query execution, interpreting analyses and coordinating responses through modular and remotely accessible software components. Finally, the last layer presents actionable information and manages interactions with external users.

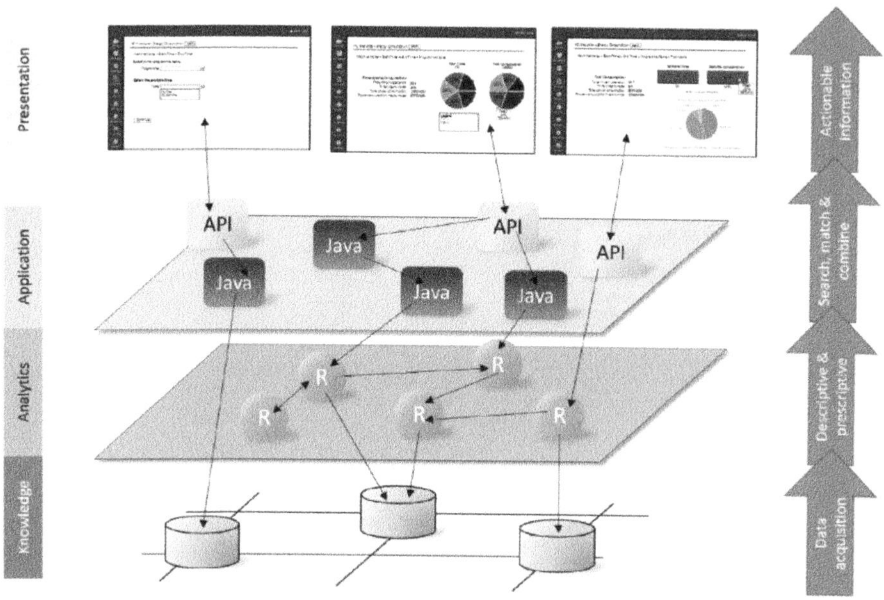

Figure 8.3. *The four layers of Mc-Analytics proposed by Terrazas et al. (2019)*

The authors experimented with this framework to analyze energy consumption data using cloud computing, thus contributing to infrastructure cost optimization. This approach provides a rigorous structure for exploiting cloud computing in a digitalized industrial context. However, adaptation is necessary before deployment, taking into account the typology of available data and the targeted optimization objectives.

Another relevant study is that conducted by Wang et al. (2017). The framework they propose is structured around three layers: the physical resource, the local server and the cloud, as shown in Figure 8.4. This framework is distinguished by a data flow connecting the workshop, the local server and cloud computing, conferring upon the cloud the function of data management. At the "physical resources" layer level, manufacturing operations or those executed by robots are referenced in the cloud layer. Similarly, raw data from cameras, sensors and controllers, mainly used for real-time monitoring, are transferred to the cloud for analysis and for potential future improvements.

The "local server" layer ensures real-time collection of machine status as well as raw data from the physical layer. Thanks to a local synchronization system, it efficiently manages the computational workload between the local server and the cloud. When computing demand is low, the local server autonomously handles task execution. In contrast, for intensive computations, such as optimization or simulation, data is aggregated by the local server and transferred to the cloud in the form of a service request.

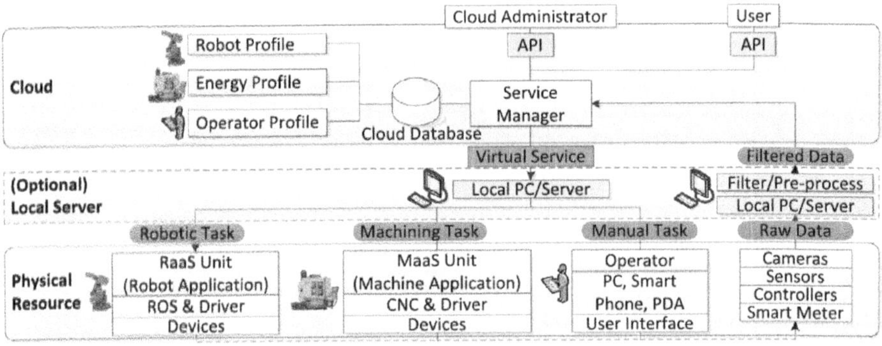

Figure 8.4. *The three layers of the framework proposed by Wang et al. (2017)*

The cloud layer coordinates and supervises all services within the production system. It integrates physical production resources and provides the necessary manufacturing services. In this layer, the service manager interacts with cloud users and executes service packages according to needs and the required computing power. In summary, the cloud system is divided into two subsystems: the first makes workshop operations accessible as services in the cloud, thus offering end users the possibility to order them; the second manages data flows between the workshop, the local server and the cloud, filtering, preprocessing and analyzing data.

In conclusion, the methodologies and frameworks proposed in the literature offer an interesting conceptual overview, but lack precision regarding deployment procedures. In some cases, data from production lines is used solely for monitoring purposes, without being exploited to improve industrial performance. Furthermore, the classification of data from the workshop is often limited to high-level categorizations, such as structured and unstructured data. A more in-depth classification is crucial to establish links between industrial needs, resources and the cloud service architecture. The following chapter presents a new detailed approach, experimented with on the ENSAM Industry 4.0 platform, Paris campus.

8.2. Integration methodology

This section presents the methodology developed by El-Helou (2023), designed to facilitate the adoption and integration of cloud computing in order to improve the performance of advanced manufacturing systems. The six phases of this methodology are detailed and structured in Figure 8.5. This result stems from an iterative and experimental process, based on an analysis of the literature and experts, as well as complementary work carried out in collaboration with engineers and doctoral students of the Industry 4.0 platform. Furthermore, the deployment framework will be addressed in section 8.3.

– Phase 1: identification of industrial objectives or needs.

– Phase 2: inspection and adaptation of the workshop's IT infrastructure.

– Phase 3: classification and categorization of industrial data.

– Phase 4: data aggregation and identification of the data lake.

– Phase 5: data analysis in the cloud and creation of dedicated processes.

– Phase 6: deployment of processes, either locally or in the cloud.

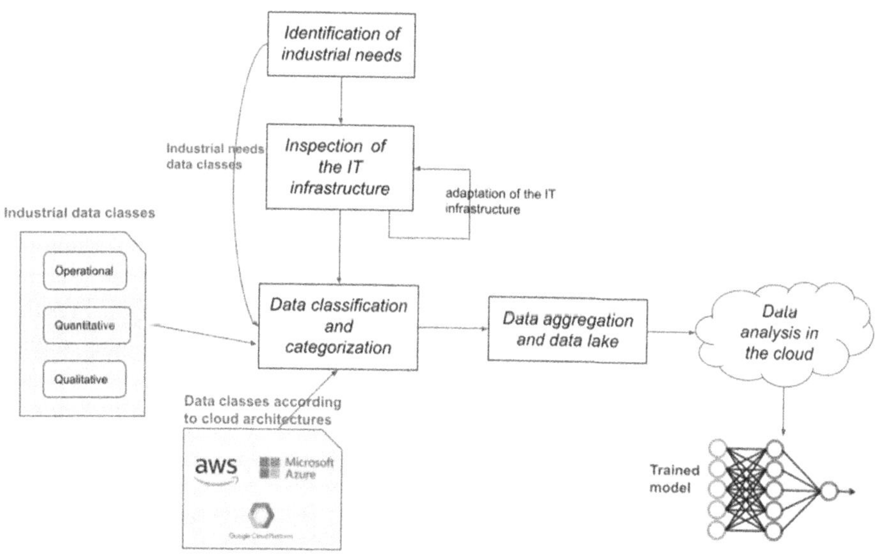

Figure 8.5. *Proposed methodology for cloud computing integration*

8.2.1. *Identification of industrial needs*

The integration of cloud computing into manufacturing processes aims to optimize functionalities such as predictive maintenance, dynamic operations control and security reinforcement. For example, in terms of predictive maintenance, the analysis of production data makes it possible to identify statistical correlations between equipment behavior and occurred degradations. These models, once trained with cloud AI resources, can monitor machine status in real time and anticipate failures, thus reducing downtime and the indirect costs associated with production loss.

For dynamic production control, digital tools and computational resources facilitate the optimization of operations scheduling in real time thanks to heuristic multi-objective resolution algorithms. These algorithms, computationally intensive, make it possible to identify a compromise that minimizes energy consumption while maximizing production cadence, as an example. Furthermore, intelligent analysis of moving data flows relative to active processes can detect anomalies and alert supervisors in the event of inconsistent activities. These applications, requiring significant analysis resources, contribute to reinforcing the security of digitalized production systems while guaranteeing their compliance with current regulations.

In conclusion, the identification of targeted objectives is the first phase of the methodology and directly influences all subsequent steps. This step is crucial for the success of cloud computing adoption. A clear definition of objectives allows companies to strategically direct their efforts. This includes the selection of appropriate technologies, the choice of cloud service providers and the integration of solutions into existing processes. By having a precise vision of what they wish to achieve, companies can also establish performance indicators to measure the effectiveness of implemented solutions. This approach promotes better resource allocation and facilitates the adjustment of strategies according to obtained results, thus ensuring a successful transition towards more advanced and digital manufacturing systems.

8.2.2. *Adaptation of the workshop's IT infrastructure*

This step is not exclusive to cloud computing deployment; the evolution of IT infrastructure is generally a prerequisite for the transition towards digitalized production systems. Its main objective is to strengthen the workshop's capacity to generate, collect and streamline data transfers. This involves the adoption of new technologies and practices that improve data management and contribute to the digitalization of production processes. The inspection of IT infrastructure includes several essential elements, such as machine connectivity, the sensor network, the IoT layer, cameras, the specification of communication protocols, the availability of local servers, as well as the complete or partial opening of the machine control layer to interact with the supervision system.

Machine connectivity is fundamental, as it allows all units to communicate with each other and with supervision systems. Effective connectivity ensures data synchronization and dynamic operations supervision. Similarly, sensors and IoT devices play a crucial role in real-time data collection. It is imperative to verify their availability and add them if necessary to meet the targeted new functionalities, notably through correct sensor installation, regular calibration and data lake sizing.

The specification of the communication protocol is also essential to ensure that all devices and systems of the IT infrastructure interact in a coherent, secure and real-time manner. It is important to define and implement standard protocols, such as MQTT, OPC UA or Modbus, according to the specific needs of the workshop. Similarly, local servers play a central role and must be evaluated in terms of storage capacity, processor performance, data security and redundancy to avoid any data loss in the event of hardware failure.

The opening of the machine control layer is a strategic aspect that depends on suppliers and the agreements established with operators. An open control layer allows flexible interaction with the digitalized environment, favoring the integration of new technologies and continuous process improvement. However, this opening must be managed with caution to guarantee the security and stability of operations.

Inspection of IT infrastructure is an iterative process, to be carried out regularly and continuously, rather than just once. Each inspection can reveal needs for adjustment or modification to align the infrastructure with the evolving objectives of the workshop, including software updates, the addition of routers, modification of communication protocols or server upgrades. In conclusion, a meticulous and continuous inspection of IT infrastructure is essential to prepare for cloud computing integration. This iterative approach ensures that the IT infrastructure remains aligned with production objectives and the requirements of a digitalized environment.

8.2.3. *Classification and categorization of data*

Industrial data is classified and separated into two different stages in order to formalize a complex link that will be useful for meeting the objectives expressed in section 8.2.1. First, this data is organized to identify a relationship between the data type and the industrial objectives. Next, the data is structured so as to be consistent with the cloud architecture and harmonized with available resources, thus facilitating its migration and exploitation in the cloud. The innovation lies in the formalization of a complex relationship between "industrial objectives" and "cloud architectures and resources" via two main stages: "industrial objectives – data type" and "data type – cloud architectures and resources". Thus, it becomes possible to link industrial objectives with the resources and architecture offered by cloud computing, as shown in Figure 8.6.

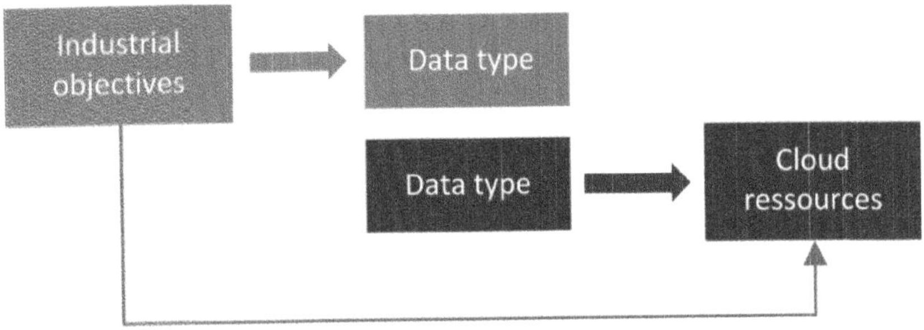

Figure 8.6. *Relationships between industrial objectives and cloud resources*

The following is organized into two sections. The first addresses the principle of data classification according to industrial objectives. The second deals with data classification according to cloud architecture and resources. These two aspects are inseparable to ensure effective integration of industrial data into a cloud environment.

8.2.3.1. *Classification aligned with industrial objectives*

This first data classification step aims to identify the type of data mainly used for each industrial need. To this end, six major industrial needs have been defined, each associated with one of three categories of industrial data: operational data, quantitative data and qualitative data.

Industrial objectives can be characterized by product and process quality, as well as all activities related to production, contributing to the satisfaction of internal and external stakeholder requirements. Similarly, industrialists primarily seek to improve their production performance in terms of cost, time and energy while taking into account environmental and societal aspects.

On the contrary, the sought-after performance can be expressed by new functionalities integrating advanced technologies and AI. Several examples can be cited, such as predictive maintenance, which has captured the attention of industrialists by enabling the development of algorithms capable of anticipating failures or performance degradation. Process visibility and product traceability are essential in sensitive industries. Virtualization, driven by the digital twin of the physical factory, is useful both in simulation and in supervision, thus optimizing production plan design time and facilitating dynamic sharing of production information. Finally, personnel and machine safety remains a priority for companies.

Thus, the three types of data identified are as follows: operational data, which includes relatively well-known information such as machine configurations, process parameters and manufacturing ranges; quantitative data, collected using IoT sensors or cameras, whose analysis makes it possible to generate information essential for performance optimization, thus enabling informed decisions based on concrete measurements; and finally, qualitative data, which aims to describe rather than measure a subject, used for subjective evaluations, such as conformity assessment or customer satisfaction. The latter are generally less structured and offer depth of analysis, allowing understanding of underlying motivations, reasoning and attitudes. However, their less structured nature poses additional challenges during their analysis.

Table 8.1 summarizes the main relationships between industrial objectives and data types. For example, to achieve an industrial objective such as optimization, operational machine data is used as input, while quantitative sensor data is used as output.

Industrial objectives	**Data type**		
	Operational data	Quantitative data	Qualitative data
Product and process quality	Input	Input	Output
Optimizations (cost/time/energy)	Input	Output	–
Predictive maintenance	Input	Input/output	–
Visibility/traceability	Input/output	Input/output	–
Virtualization	Input	Input/output	Input/output
Safety/security	–	Input	–

Table 8.1. *Relationship between industrial objectives and data type*

8.2.3.2. *Classification aligned with cloud architectures and resources*

This classification approach was proposed after a careful examination of the architectures and resources of the most widely used clouds, such as AWS from Amazon, Azure from Microsoft and GCP from Google. The linking between industrial data and cloud resources depends on the typology of collected data, thus allowing each data type to be linked to a specific resource or group of resources in the cloud. The objective is to facilitate both data migration and analysis. Four key typologies of industrial data have been identified:

– Photos/videos: images and videos collected in an industrial context come from various industrial cameras. These include supervision cameras, guidance cameras for robots and counting cameras installed at stock locations. This data is integrated into cloud cognitive services to be analyzed, thus generating collision prevention models, for example.

– Machine data: this type of data comes from machine control systems and robot control cabinets. It includes, among others, alarms, machine operating status and robot positions. This data is integrated into ML services in the cloud in order to extract hidden information, such as the detection of patterns that could indicate premature wear, thus enabling preventive inspection planning.

– IoT and sensors: this category encompasses all wired and Wi-Fi sensors installed on manufacturing equipment or at production lines, enabling the collection of data such as temperature, acoustics, vibrations, energy consumption, etc. This data is integrated into the IoT analysis service in the cloud. The analysis can run in real time thanks to models detecting any deviation, or in deferred mode for in-depth analysis of archived data, in order to extract complex phenomenological correlation models.

– Operations data: this includes all data related to the production plan being executed at the workshop level, such as machine programs, geometric models of parts to be manufactured, quality repositories, equipment configuration and operating parameters, as well as information for production supervision. Operations data is collected and stored in cloud storage to be used later in a context extended to other data types.

Considering the resources available in the different clouds, Table 8.2 gives an overview of the possible links between the data typologies identified in a digitalized production systems context and the specific resources available. This confirms that the proposed approach is consistent with the different cloud computing solution providers.

8.2.4. *Data aggregation and data lake construction*

After the classification of industrial data, this data must be collected and aggregated in local storage while taking into account their timestamps. This makes it possible to obtain a reliable and meaningful overview of the data as well as their provenance. For example, in the context of predictive maintenance, timestamps of data from machines and sensors are essential. It is therefore crucial to collect and aggregate this data in a structured manner in the data lake, preserving their temporality to facilitate their analysis. By combining classification and timestamp-based aggregation, this approach makes it possible to establish precise correlations and monitor the evolution of events through in-depth analyses.

	Type of industrial data	**Associated cloud resources**	**Azure**	**AWS**	**Google Cloud platform**
1	Photos/videos	Cognitive services	Custom Vision	Rekognition Polly	Cloud Vision
2	Raw machine data	Machine learning services	Azure Machine Learning	SageMaker AML	Cloud Machine Learning Engine
3	IoT and sensors	IoT analysis or ML services	IoT Hub IoT Edge	IoT Core Greengrass	Cloud IoT Core (Beta)
4	Operations related data	Cloud storage	Blob Storage Data Lake Store	Simple Storage Service (S3) Elastic Block Storage (EBS)	Cloud Storage Cloud Datastore

Table 8.2. *Associated cloud resources for each data type*

The construction of a well-structured data lake is a prerequisite for establishing the relationship between industrial objectives and cloud architecture. It is a modern data storage concept that allows large quantities of data to be kept in their raw and original format. Unlike traditional data warehouses, which structure data into hierarchical tables, data lakes adopt a flat architecture. This approach allows data to be stored as files or object storage, often including the data time frame.

One of the main objectives of a data lake is to provide increased flexibility in data supervision and use. By allowing the storage of structured, semi-structured and unstructured data, a data lake can adapt to a wide variety of data types, ranging from text files and images to data streams from IoT sensors. Storing data in their raw format allows analysts to process and transform them according to their specific needs, without loss of initial information. By offering access to a vast quantity of archived and real-time data, data lakes facilitate the extraction of knowledge and valuable information through intelligent computing tools, including big data analytics and ML. The data lake relies on a flat architecture that allows direct and simple access to data, without having to go through complex hierarchical structures. This simplifies data manipulation and search, making information extraction faster and more efficient.

In the experiment presented here, industrial data from various sources such as cameras, machines or manufacturing execution systems (MES), are centralized in the data lake. This process is automated by scripts that ensure a continuous flow of data from their source to their storage in the local data lake. These scripts often integrate validation and cleaning mechanisms to guarantee data quality and integrity. Furthermore, professional solutions offering an integration platform as a service (iPaaS), enabling the connection of applications, data and APIs in cloud, on-premise or hybrid environments, can be used to deploy a data lake.

To set up a data lake intended for the use of industrial data, several prerequisites are necessary. First, the storage infrastructure must be capable of adapting to constantly increasing data volumes. High availability and redundancy are essential to ensure data continuity and prevent losses. It is crucial to connect all industrial data sources, including IoT sensors, MES, cameras and machines. Automation of data flows using scripts or ETL (Extract, Transform, Load) tools facilitates data transfer from sources to the data lake.

Furthermore, data validation and cleaning processes are indispensable to guarantee the accuracy and reliability of stored information. Data categorization facilitates information search and access, making the data lake more efficient. Data security is also a priority, requiring robust measures such as encryption of data in transit and at rest, as well as access management. It is also important to ensure that the data lake complies with local and international regulations regarding data privacy and protection.

From an economic perspective, the global data lake market could reach a value of approximately \$90 billion USD by 2032, representing a significant increase compared to its value of \$13.7 billion USD in 2022. This growth will be characterized by a compound annual growth rate (CAGR) of 21.3% for the period from 2023 to 2032. This study was published in March 2024 by Market.us (Market.us 2024).

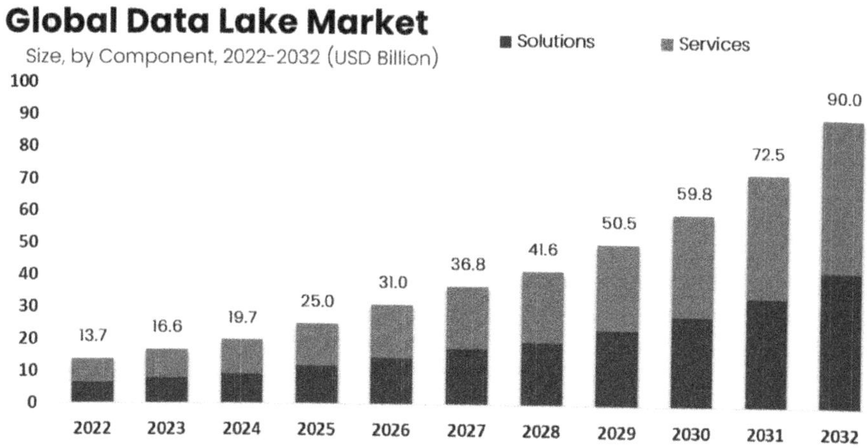

Figure 8.7. *Global data lake market. For a color version of this figure, see www.iste.co.uk/benfriha/industrialprocesses.zip*

In summary, an organized data lake enables the efficient management of large quantities of diverse data from industrial environments and facilitates its integration with cloud computing solutions. However, several factors can limit the use of data lakes, including the complexity of their deployment, the lack of specialized skills, high costs, particularly for small and medium-sized enterprises, and cybersecurity concerns.

8.2.5. *Analysis in the cloud and creation of specific processes*

After data aggregation and configuration of a local data lake, the data is ready for migration to the cloud to be analyzed according to its use and type. Each data typology migrates to the appropriate cloud resource, in order to be compatible with its architecture. Data migration can be performed manually by uploading data to the cloud or in an automated manner using scripts developed in various programming languages. It is this latter approach that was deployed for the experiment.

Once the data is migrated to the cloud, its analysis can begin. To this end, the production supervisor must first define the data pipeline according to the industrial purpose and analysis parameters. This includes the definition of key metrics, performance criteria and final analysis objectives.

The data pipeline can be compared to a computational algorithm aimed at ensuring an efficient processing flow. The first phase of this process consists of selecting data in accordance with the analysis objectives. Next, data preprocessing intervenes, including actions such as cleaning to eliminate outliers and correct errors, normalization to make data comparable and enrichment to add additional information necessary for analysis. Furthermore, certain data may be adapted to analysis models, which may include the creation of new variables or characteristics from existing data.

The analysis phase follows, involving the application of analytical methods such as statistical analysis, ML or other advanced techniques to extract insights and models from data. Once the analysis process is complete, interpretation of the results allows understanding of their scope and industrial implications. This step may include result visualization, report generation and communication of conclusions to stakeholders. The result can also translate into a trained computational model, then repatriated to the factory servers for local exploitation, notably for anomaly detection, collision prevention or other specific applications.

Finally, the objective is to transform generated data into actionable knowledge. This can translate into recommendations for industrial process optimization, detection of new opportunities or improvement of decision-making. By following these methodical steps, companies can valorize their data, obtain valuable insights and strengthen the efficiency of their industrial operations.

In general, the data analysis process in the cloud is simple, fast and does not require specific computer skills because a set of integrated resources and tools is available on a pay-as-you-go basis.

8.2.6. *Deployment of the methodology*

In general, cloud computing can produce three levels of results according to expressed industrial needs and the deployment context. The first level is offline analysis, where data is processed without real-time interaction with ongoing operations. This type of analysis is ideal for retrospective reporting and the development of phenomenological models based on the interpretation of historical data, requiring significant computational resources. The results obtained can then be used for prediction, optimization and intelligent supervision of operations.

The second level consists of analyzing data without producing direct actions on production systems. Production data is updated and analyzed in real time at the cloud level, enabling dynamic monitoring and visualization of performance indicators. This offers an overview of the production line status and operations progress. Thus, the multimodal interactive interface of the digital twin becomes dynamic, and its portability allows the sharing of indicators and their evolution among different internal or external stakeholders, for example. This information allows decision-makers to quickly understand trends and react to anomalies or opportunities as they arise.

The third level is online analysis with feedback actions on the industrial system. In this configuration, analytical models receive and analyze data in real time and can suggest actions, such as adjustments or changes of variables, that the MES will apply to the industrial system. For example, analytical models can suggest a production slowdown if the analysis of upstream data reveals deviations at the limit of the quality tolerances defined in the repository. This level confers upon the MES increased autonomy and intelligent control of industrial processes, thus contributing to improving overall performance.

Concerning this deployment phase, it aims to implement the trained computational models encapsulated in an application responding to a specific industrial need, thus corresponding to the first level of results. To do this, relevant data is identified and sent to the cloud for analysis, making it possible to produce a computational model that represents the interactions between the different variables and collected data. Once validated and deemed sufficiently precise, these models can also be reused or integrated into other comparable industrial applications.

These applications can be deployed in two ways. They can remain in the cloud, thus benefiting from the computing power and scalability of cloud computing, which facilitates updates and continuous maintenance. Alternatively, applications can be downloaded and used locally, which is particularly useful in environments with limited connectivity or for data security reasons.

Deployment of applications in the cloud is easier and requires fewer specialized skills. However, some companies prefer local deployment over remote deployment for reasons of sensitive data security. The deployment procedure, whether in the cloud or locally, depends mainly on the level of data security and also on the availability of IT infrastructure and local resources. Furthermore, industrial needs requiring online output with feedback actions on the industrial system must be analyzed according to the risk of non-response, linked to a connectivity problem, for example. In this case, edge or local deployment will be preferred.

8.3. Deployment framework

The proposed framework is designed to implement the methodology described in section 8.2. Figure 8.8 shows its integration into the Industry 4.0 platform architecture, addressed in section 5.2 of Chapter 5. This framework is distributed into three main spaces, connected by an information and data flow. The first space covers all operational perimeters where data is generated and their sources identified, corresponding to zone 1 of the Industry 4.0 platform. The second is dedicated to the preprocessing and local storage of collected data, integrated into zone 3 of the platform, which is dedicated to operations supervision and control. Finally, the cloud computing space is that to which data is transmitted for analysis and generation of computational models, responding to optimization objectives or decision-making processes. This configuration enables a rigorous structuring of the collection, classification, storage, migration, analysis and redeployment process of models in production workshops.

Thus, the first space encompasses all workshop entities, such as machines, sensors, cameras, as well as the MES, which generates data contextualized according to the specific conditions of their collection. For example, a vibration detected by a sensor must be enriched with complementary information, such as the material being manufactured, the tool used and the cutting conditions.

The second space is dedicated to data produced in the workshop, which is then historized, aggregated and classified in the data lake according to their source, their typology and their timeframe capture. To do this, specific scripts must be developed for each data type and each source, taking into account their intended use. Edge computing and the data lake offer optimal solutions for data storage and preprocessing before their migration to the cloud.

Data migration from the data lake to the cloud is then carried out by scripts that can be activated manually or operate autonomously. These scripts are adapted to the selected cloud computing service provider, each provider having its own approach to data migration. Finally, most cloud service providers allow the local repatriation of trained and validated computational models, in order to meet the specific requirements of the company.

The third space is dedicated to cloud computing, where each data type is associated with a specific cloud resource, according to its typology and intended use. It is in this environment that data analysis is carried out and that preconfigured models are trained using data generated by the factory. To encourage the adoption of cloud computing solutions, providers have developed attractive marketing strategies, offering ready-to-use computational modules, notably for image processing and IoT analysis. Furthermore, they innovate by regularly updating their offerings, which makes cloud computing simpler and more accessible to use.

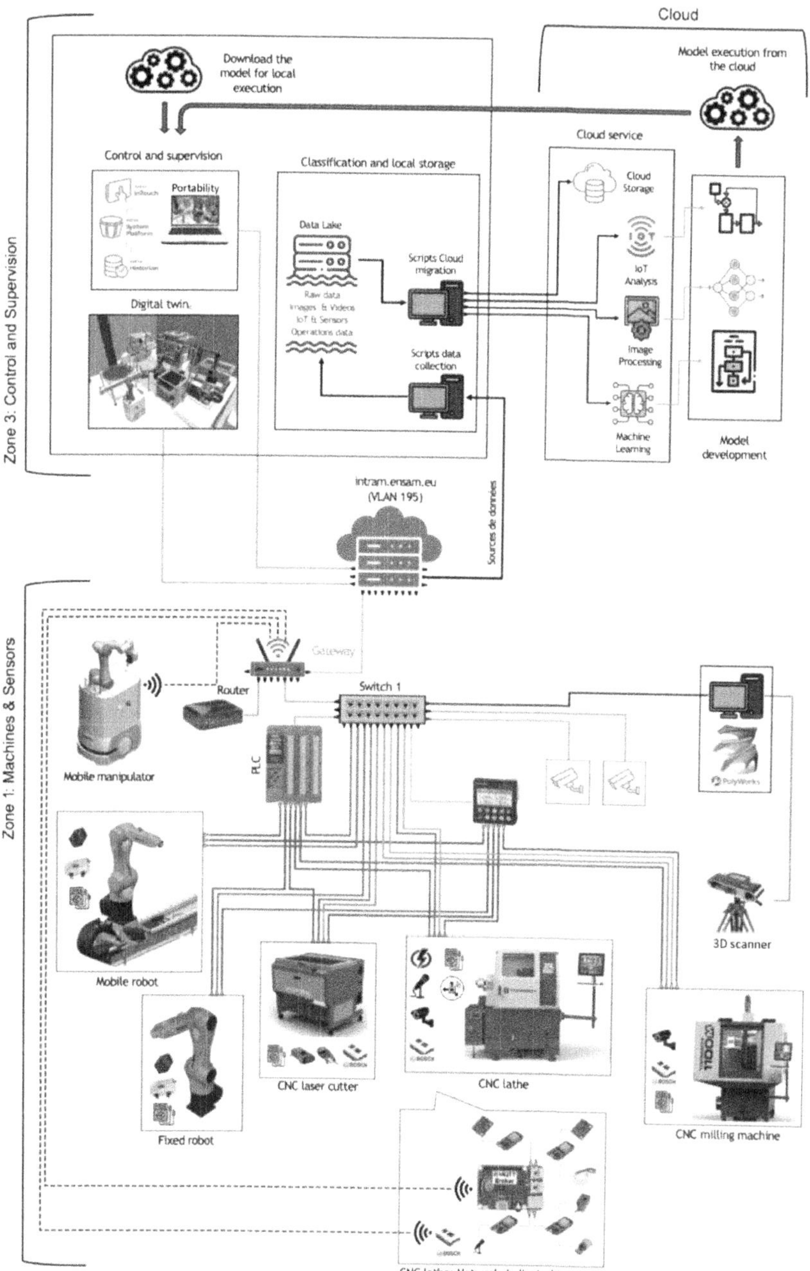

Figure 8.8. *Cloud computing deployment framework. For a color version of this figure, see www.iste.co.uk/benfriha/industrialprocesses.zip*

In addition, clouds offer tools and an environment allowing the development of customized algorithms, integrating specific computational modules as well as a data pipeline. Depending on the level of digital maturity of companies and the complexity of the phenomenological behaviors to be modeled, a combination of preconfigured and customized modules often proves indispensable to achieve the desired objectives.

The proposed configuration structures the implementation, deployment and progressive execution of the cloud computing integration methodology in an industrial context. To this end, the Industry 4.0 platform was used as an experimental tool, enabling the generation of necessary data and the execution of several tests simultaneously. Chapter 9, which addresses the "optimization" functionality, presents a cloud computing use case illustrating a multi-objective optimization need considered complex. This case aims to determine the cutting conditions, during a machining operation, that minimize energy consumption, machining time and vibratory disturbances.

8.4. Synthesis

The proposed approach begins with the definition of industrial objectives, involving a clear understanding of expected results, such as improving product quality, optimizing production processes or reducing operational costs. Next, the factory's IT infrastructure is inspected to evaluate its computing, storage and data collection capacity, including network verification.

The following phase concerns the local collection and classification of industrial data before their transmission to the cloud for analysis. The cloud offers computing power and advanced analytical tools, enabling the generation of adapted models, which can be deployed either locally in the factory or in the cloud, depending on specific needs and security constraints. These models contribute to improving industrial performance through their optimization capabilities, thus reinforcing the MES in its role of autonomous and intelligent supervision.

To implement these principles, a framework was developed and experimented with at the Arts et Métiers Industry 4.0 platform, illustrating the effectiveness of this approach. This framework demonstrated the added value that cloud computing resources can bring to the optimization process of industrial systems. However, it also highlights the difficulties encountered when designing an algorithm integrating preconfigured and customized models for complex analyses, often multi-objective and multi-parametric.

In conclusion, this chapter presents an experimental approach enabling a structured and progressive integration of the new capabilities offered by cloud computing into the manufacturing processes of a digitalized production workshop. This approach stimulates innovation in the manufacturing sector, thus contributing to its digital transformation.

9

Digital Optimization

Digital optimization is an emerging concept that consists of using data collected by sensors or cyber-physical industrial systems to improve industrial processes. This data, analyzed using artificial intelligence (AI) tools, can reveal phenomenological correlations whose interpretation can lead to optimization ideas. Furthermore, this information can influence other phases of a product's lifecycle. For example, the measured energy profile, which encompasses all production operations of a component, can identify those that are most energy-intensive. This information can be exploited to optimize not only production but also component design. Thus, for production, the Manufacturing Execution System (MES) could schedule energy-intensive operations during periods when the energy consumed is decarbonized, for example, when the energy produced by photovoltaic panels installed on the factory roof is maximal. Regarding design, engineers could adjust the morphology of certain parts of the component in order to reduce its energy consumption during manufacturing. These examples show that the digitalization of production processes opens the way to new optimization opportunities, including all phases of a product's lifecycle.

This chapter presents a concrete experiment in digital optimization, using generated data to improve production processes. This experiment was carried out on an Industry 4.0 platform, requiring the development of an innovative algorithm including cloud computing tools. The objective is to determine the machining parameters of a CNC lathe, such as rotational speed, depth of cut and cutting tool feed speed, that minimize energy consumption, machining time and vibratory footprint. This multi-objective approach is made possible thanks to the digitalization of the machining process, facilitated by the installation of vibratory and energy sensors, which made it possible to collect a vast set of data relating to the manufacturing of a specific part.

9.1. Introduction to the problem

Before addressing the characterization of the optimization problem studied here, a brief introduction will trace the evolution of improvement and optimization principles through the industrial revolutions.

9.1.1. *Evolution of industrial optimization*

The optimization of industrial processes has evolved over the centuries, marking key stages in the history of production and having a profound impact on working conditions. In the pre-industrial era, optimization methods were mainly based on craft practices aimed at improving the efficiency of manual labor. However, working conditions were often arduous, with long hours and little safety for workers.

The First Industrial Revolution radically transformed production through mechanization. Although this made it possible to increase productivity, the early factories were often synonymous with difficult working conditions, characterized by repetitive tasks, rhythms imposed by machines and few protections for workers.

At the turn of the 20th century, theorists such as Frederick Taylor introduced the principles of scientific management. Taylorism, by seeking to increase work efficiency, led to a fragmentation of tasks and increased specialization, thus reducing worker autonomy while increasing productivity. Although this approach benefited production, it is often criticized for having made work more monotonous and alienating. In parallel, reflections on working conditions emerged, with efforts to reduce fatigue thanks to the motion studies of Frank and Lillian Gilbreth.

The 20th century saw the emergence of systemic methods, such as Henry Ford's assembly lines, which facilitated mass production but intensified repetitive work. However, improvements began to appear, notably with the development of Lean Manufacturing at Toyota, which emphasizes continuous improvement and increased worker involvement in the production process. The Toyota model also introduced practices aimed at reducing waste of time and effort.

At the end of the 20th century, the advent of computerized systems and automation further modified working conditions. The automation of repetitive tasks reduced the arduousness of certain functions, although it raised new questions concerning the reduction of manual jobs. Robotization, while improving precision and efficiency, progressively replaced certain human tasks, thus contributing to enhancing safety by removing workers from high-risk environments. However, it also led to a polarization of employment: on the one hand, a decrease in low-skilled positions, and on the other hand, an increase in demand for specialized technical skills.

Industry 4.0, in the 21st century, brings a digital transformation of industrial processes, integrating technologies such as the Internet of Things (IoT), artificial intelligence (AI) and digital twins. These innovations make production smarter, more flexible and more reactive, while having an impact on working conditions. Work becomes more focused on data management and interpretation, offering opportunities for improving working conditions through increased safety monitoring, reduction of physically demanding tasks and more ergonomic work environments. This transformation thus shifts sought-after skills towards specialized levels of expertise, forcing workers to retrain to adapt to this new industrial landscape.

9.1.2. *Characterization of the problem*

The objective is to determine the optimal parameters for a machining operation on a computer numerical control machine of the TORMACH 15L SLANT-PRO CNC lathe type, using a conventional material removal process. This machine allows various turning operations such as longitudinal turning, facing, contouring, threading and parting off. To do this, a series of sensors was installed to collect physical machine data during the execution of manufacturing sequences, carried out within the framework of the research work of Hakam (2024).

To carry out this experiment, the "facing" operation was selected due to its reduced execution time. This operation was repeated 88 times, with variations in cutting conditions, to collect a sufficient volume of data for modeling. It is characterized as follows: a raw cylinder of 25 mm diameter and 50 mm length is held in the chuck over a depth of 20 mm, then a facing of 6 mm is executed, as shown in Figure 9.1.

The system is subjected to the variation of three input parameters: cutting speed, depth of cut and feed per revolution. For each set of input parameters, three output data are measured: execution time, energy consumption and generated vibrations. This configuration was chosen due to the evident phenomenological relationships between input parameters and output data. For example, an increase in feed per revolution would lead to an increase in energy consumption while reducing execution time. This underscores the importance of a business analysis of industrial processes in any optimization approach.

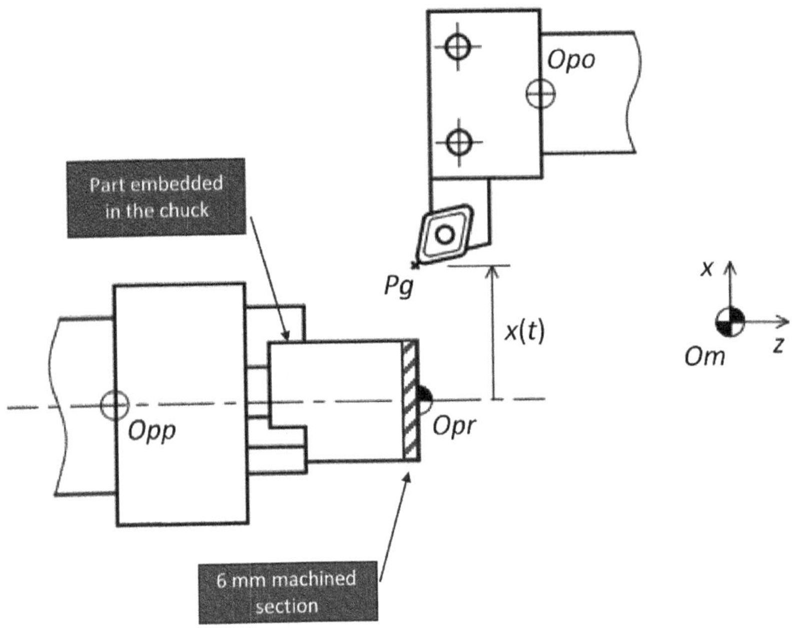

Figure 9.1. *Facing-type machining operation*

9.2. Optimization phases

The optimization approach consists of four main phases. The first phase concerns the modeling of the system, considering input data, output data and objective functions. The second phase consists of designing an adapted infrastructure, defining the needs for sensors, connectors and software. The third phase aims to create an analysis algorithm that maximally exploits the computational resources predefined in cloud computing. Finally, the last phase consists of verifying the consistency of results and interpreting trends. This approach is iterative, facilitating progressive adjustments in order to improve results, as shown in Figure 9.2.

To validate the relevance of this approach, the experiment is based on a multi-parametric and multi-objective optimization, carried out from a set of 88 machined parts. Three manufacturing parameters were varied, and for each part, three output data were collected.

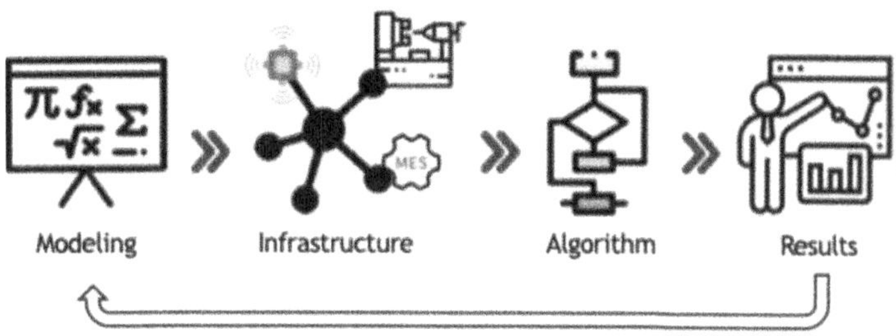

Figure 9.2. *Optimization approach*

9.2.1. *Modeling*

As a reminder, during an elementary manufacturing sequence, input parameters define the cutting conditions, including cutting speed, machining depth and axial tool feed. Output data includes physical measurements such as time, generated vibrations and energy consumed. This data, once transmitted and stored in the cloud, will be analyzed to reveal phenomenological correlations and establish a computational model linking input parameters to observed results, as shown in Figure 9.3.

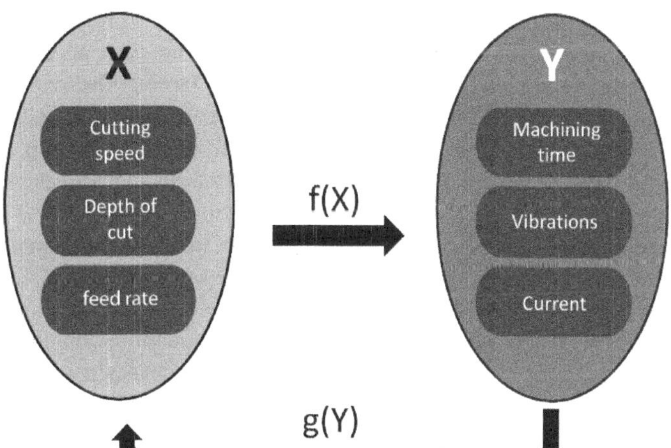

Figure 9.3. *Representation of the passage from space X to space Y*

The function "f" represents this model, which, from a set of X data, makes it possible to predict a set of Y data. This multi-objective optimization approach is developed using tools available in cloud computing. The data pipeline tool makes it possible to structure the processing flow and visualize the entire algorithm, including both intelligent and conventional calculation blocks at precise stages. The use of cloud resources, notably artificial intelligence blocks, is compatible with the multi-physical and multi-parametric complexity of the system to be solved.

However, the use of AI tools in industrial processes is subject to prior verifications that confirm, for example, the existence of the projection function Y = f(X). This means that there is a certain correlation, validated by business experts, between the two considered sets. For example, if the depth of cut increases, energy consumption increases proportionally. In contrast, execution time decreases inversely proportionally. This type of relationship, which expresses the influence of input data variation on measured output data, is decisive for the success of the approach.

The set of input data (set X) includes cutting speed (Vc), expressed in meters per minute (m/min). This parameter is used to calculate the workpiece rotational speed, in revolutions per minute (rpm), according to the following formula:

$$N = \frac{1{,}000 \times Vc}{\pi \times d} \quad [9.1]$$

It also includes depth of cut, which represents the thickness machined during each pass of the tool, expressed in millimeters (mm). Furthermore, feed per revolution corresponds to the distance traveled by the tool at each complete revolution of the workpiece, expressed in mm/rev. During the experiment, the machine's input parameters vary. Cutting speed oscillates between 100 and 200 m/min, with an increment of 10 m/min. Depth of cut ranges from 0.5 to 2.0 mm, with a step of 0.5 mm. As for feed per revolution, it takes only two distinct values: 0.1 mm/rev and 0.4 mm/rev. This combination of input parameters is applied to the machining of 88 parts, each being carried out with a unique set of machining parameters.

In contrast, the set of output data (set Y) consists of the physical quantities measured for each of the 88 machined parts. These quantities include machining time for each operation, expressed in seconds, as well as energy parameters such as instantaneous voltage and current, expressed in volts and amperes, which make it possible to deduce power and, consequently, energy consumed. Chuck vibrations are also measured, expressed in meters per second squared (m/s^2) in the three spatial directions. For voltage, current and vibrations, the root mean square (RMS) – or effective value – will be calculated over the duration of each experiment. This approach is preferred over instantaneous value because the average value makes it possible to determine whether the machine is heavily solicited during the operation, thus exceeding the proper functioning threshold.

9.2.2. *Data and IT infrastructure*

Before detailing the infrastructure necessary for the machining parameter optimization project, it is essential to integrate the repositories in which data will be classified from the earliest phases of the project.

The modeling section thus highlighted two sets of data: input data and output data. Due to the nature of the project, notably using cloud computing, two repositories can be distinguished: that of the Industry 4.0 platform and that specific to cloud computing resources, where the analysis will be carried out.

Referring to Table 5.1 (see section 5.3.1) for the classification of data within the Industry 4.0 platform, it appears that process optimization relies on the use of three types of data: machine data, operations-related data, as well as data from sensors.

Machine data concerns, in particular, operation timestamps. Operational data encompasses input information related to operation execution, such as cutting speed, depth of cut and feed per revolution. As for sensor data, it constitutes output data, exploited either for the modeling of phenomenological correlations in deferred time, or for other applications such as deviation detection, in real time.

Furthermore, data classification is also influenced by cloud resources, depending on the selected provider. Most digital platforms offer a varied range of services that require prior data aggregation. For this experiment, the Microsoft "Azure Cloud" platform was chosen due to its flexibility and the available tools. Thus, the two datasets are associated with Cloud Storage and Azure Machine Learning resources.

Regarding infrastructure, the starting point is based on the characteristics of the CNC machine, connected to the production line network, as shown in Figure 5.6. It is then essential to define the IoT device compatible with this machine and aligned with the targeted optimization objectives. This IoT device consists of sensors, controllers and an IT architecture enabling data collection, aggregation and transmission.

To this end, several configurations were tested. The final configuration retained, shown in Figure 9.4, is distinguished by its ability to transmit data quickly using the MQTT protocol. Each sensor is connected to an ESP32 card, which acquires measurements and makes them available on a Raspberry Pi card via the MQTT Broker. The data is then transmitted, hosted and analyzed using the tools available on the Azure Cloud platform.

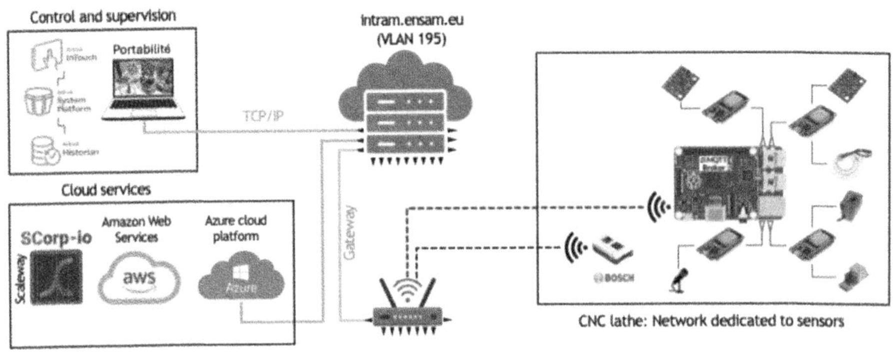

Figure 9.4. *Local architecture*

It should be noted that the Bosch sensor sends data directly to the AWS broker via the local Wi-Fi network and the virtual local area network (VLAN), in accordance with the diagram imposed by the manufacturer. Finally, interference induced by data transmission is reduced by insulated cables.

9.2.3. *Data analysis algorithm in the cloud*

The analysis algorithm is developed with Microsoft Machine Learning Studio, which simplifies the creation of a pipeline to structure the data flow and processing steps. At each step, it is possible to integrate machine learning modules, including options predefined by the cloud platform, such as neural network regression, as well as custom modules. Figure 9.5 presents the data pipeline designed specifically for this experiment.

Data collected from sensors installed on the CNC lathe is first normalized between 0 and 1. This normalization is often essential in machine learning to bring inputs to similar orders of magnitude, thus facilitating algorithm convergence. This step is integrated into the neural network, which calculates a regression to estimate output values as a function of input parameters, in other words, to determine the projection function f(X). This regression algorithm is a predefined module in Azure, which can be used directly by specifying network parameters, such as inputs, outputs, number of layers and number of nodes.

For this experiment, the optimal input parameters are those that minimize the objective functions relating to execution time, generated vibrations and energy consumption. At the end of the calculation process, the optimal machining points will be identified on the Pareto front, which represents the set of optimal solutions. To select the optimal solution from this set, the L2 norm, also known as the "Euclidean norm", is applied. Consequently, the point with the smallest norm is considered the optimal point.

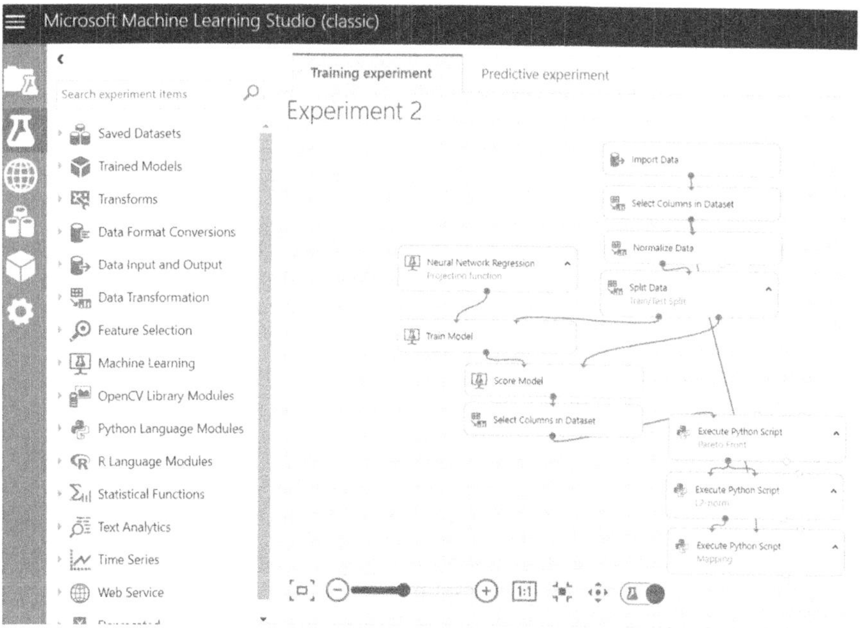

Figure 9.5. *Algorithmic pipeline in Microsoft Machine Learning Studio*

To obtain the points of the Pareto set in a three-objective problem, the following equation is applied, where objective 1 (Obj_1) corresponds to vibrations, objective 2 (Obj_2) to energy and objective 3 (Obj_3) to machining time:

$$J = w_1 * obj_1 + w_2 * obj_2 + (1 - w_1 - w_2) * obj_3 \quad [9.2]$$

The L2-norm function makes it possible to find the distance between the origin and each point of the Pareto set by applying the following formula:

$$L_2\, norm = \sqrt{(x - x_o)^2 + (y - y_o)^2 + (z - z_o)^2} \quad [9.3]$$

The minimization of this distance makes it possible to find the Pareto point that corresponds to a balance between the objective functions.

The Pareto front and the L2-norm method are not available in Azure as predefined resources. Thus, a Python script was created and integrated into the pipeline.

Once the global minimum of the projection function is identified, a return to the input space makes it possible to determine the input parameters associated with this global minimum. This makes it possible to identify the optimal parameters that simultaneously minimize execution time, energy consumption and machine vibrations, considering the priorities defined for each objective, expressed by the weights w_i. These weights adjust the relative importance of each objective, enabling optimization adapted to the specific requirements of the process.

The result of this step is a trained algorithm hosted in the cloud, ready to be deployed directly from the platform or downloaded for use at the Industry 4.0 platform. However, effective local exploitation requires specific skills, notably to adapt it to similar applications, such as the optimization of machining parameters for operations such as longitudinal turning, chamfering, parting off or threading.

9.2.4. *Results and interpretation*

According to the data pipeline presented in Figure 9.5, a three-dimensional surface is obtained, modeling all possible solutions that minimize time, energy consumption and vibrations, as shown in Figure 9.6.

However, before proceeding with the optimization of input parameters, it is essential to verify the consistency of the obtained results. To do this, a 4D projection is performed, making it possible to visualize the behavior of each output parameter as a function of the three input variables.

For example, consider the "time" output parameter: machining time is represented as a function of the three input parameters, as shown in Figure 9.7. Depth of cut and cutting speed are arranged on the horizontal plane, while the output quantity, here "time", is placed on the vertical axis. The third variable, feed per revolution, is illustrated by curves of different colors: the red curve represents a feed of 0.1 mm/rev, and the purple curve corresponds to a feed of 0.4 mm/rev.

The impact of all input parameters on machining time is clearly perceptible, which is coherent because increasing cutting speed, depth of cut or feed per revolution inevitably leads to a reduction in machining time.

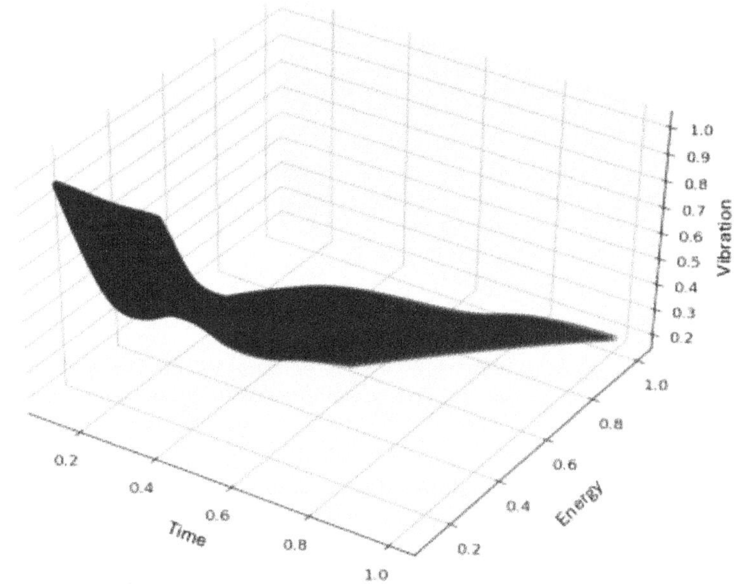

Figure 9.6. *Results projected in 3D*

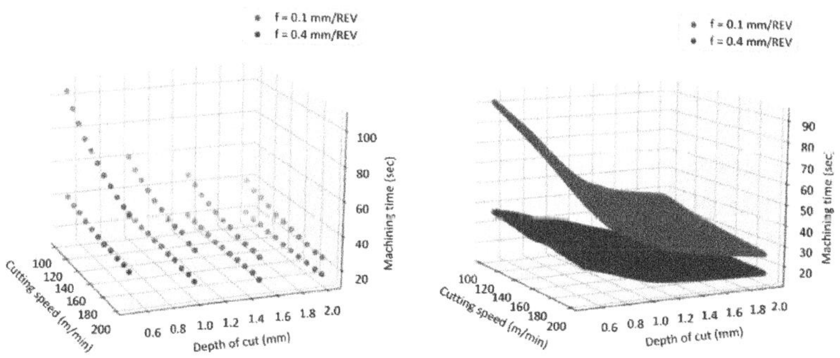

Figure 9.7. *Machining time as a function of input parameters. For a color version of this figure, see www.iste.co.uk/benfriha/industrialprocesses.zip*

Regarding the "vibration" output parameter, Figure 9.8 presents the RMS of vibrations as a function of the input parameters. As already mentioned, the average vibration makes it possible to evaluate whether, during the operation, the lathe is heavily solicited, particularly beyond a reasonable threshold to ensure its proper functioning.

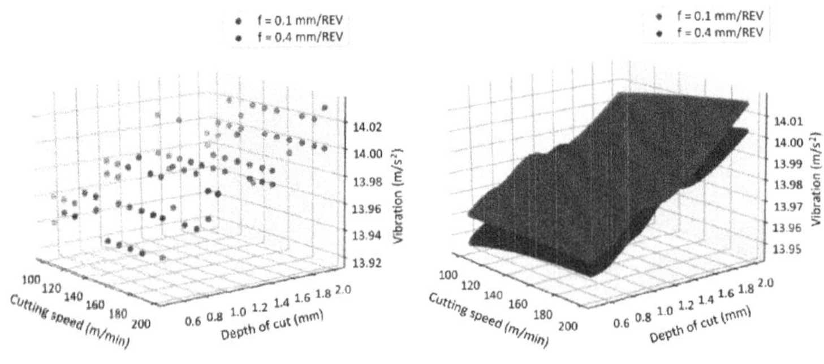

Figure 9.8. *Vibrations as a function of input parameters. For a color version of this figure, see www.iste.co.uk/benfriha/industrialprocesses.zip*

The graphical representation shows that machine vibrations vary slightly as a function of input parameters (between 13.95 and 14 m/s^2). This result is explained by the material used, the polymer Polyoxymethylene (POM), which does not require significant cutting force, thus reducing vibrations during machining, unlike harder materials.

For the third output parameter, "energy consumption", the calculation is based on the following formula:

$$\mathrm{E(KWh)} = \frac{1}{1{,}000}\int_0^T U(t) * I(t) * dt \qquad [9.4]$$

U and I denote respectively the voltage in volts and current in amperes, and T represents the machining time in hours. The machine being connected to a French single-phase power supply, its input voltage remains constant at 230 V and 50 Hz. Only the effective current increases when cutting parameters are increased. However, this increase in current is solely related to usage time and not to an instantaneous variation of current during machine operation. As shown in Figure 9.9, the energy consumption curves follow the same trends as those of machining time.

Regarding the final steps of the optimization process, the solutions obtained, visualized as a blue surface in a 3D graph, are presented in Figure 9.6. These solutions are then projected into 2D graphs, making it possible to compare objectives in pairs, as shown in Figure 9.10. In these graphs, the Pareto front is represented in red, while the point corresponding to the L2 norm, indicating the minimum distance between the origin and the Pareto front, is marked in orange.

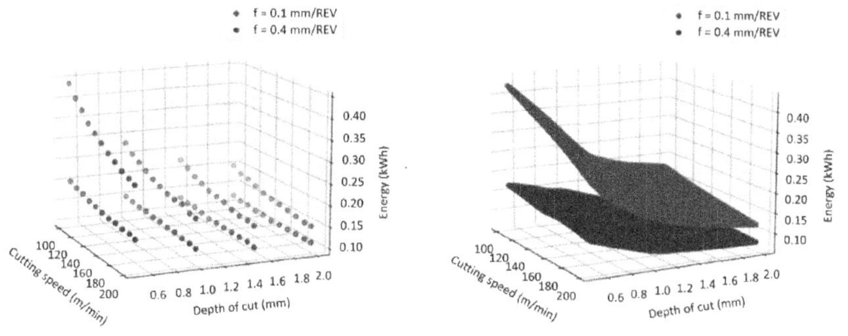

Figure 9.9. *Energy as a function of input parameters. For a color version of this figure, see www.iste.co.uk/benfriha/industrialprocesses.zip*

Finally, three potential solutions are identified, requiring a choice. First, the graph representing the relationship between vibrations and time, shown in Figure 9.10b, presents similarities with that between vibrations and energy, visible in Figure 9.10c. This leads to identical values for the machine's cutting parameters. Thus, two solutions remain to be considered: that from the energy-time projection, presented in Figure 9.10a, and that from the vibration-time projection.

The energy-time projection highlights a linear relationship, with a Pareto front confined to a single point corresponding to a minimum of both time and energy consumption, designated by point "a" on the curve.

Consequently, the optimal solution is that which minimizes the distance between the origin and the Pareto front, indicated in red, as shown by the orange straight line in Figure 9.10b and 9.10c.

The L2-norm value in the "vibration-time" projection is Vib = 13.95 m/s^2 and T = 27.19 s. Similarly, the L2-norm value in the "vibration-energy" projection is Vib = 13.954 m/s^2 and E = 1.8099 kWh.

The projection of these two values into the input space X gives the optimal input parameters, corresponding to:

– cutting speed = 198 m/min;

– depth of cut = 0.59 mm;

– feed per revolution = 0.4 mm/rev.

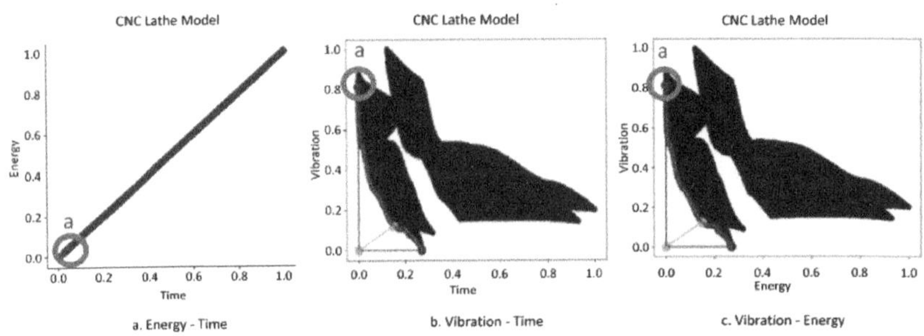

Figure 9.10. *Plotting results in 2D graphs. For a color version of this figure, see www.iste.co.uk/benfriha/industrialprocesses.zip*

These input parameters correspond to the machine's cutting parameters that offer a balance in terms of optimization priority between machining time, vibrations and machine energy consumption. This translates into the following weights:

– $w_1 = 50\%$;

– $w_2 = 0\%$;

– $w_3 = 50\%$.

The value $w_2 = 0\%$ is explained by the absence of variation in energy consumption during the operation. Thus, the algorithm makes it possible to modulate the weighting of optimization objectives, offering production supervisors the possibility of defining specific priorities for each output parameter. For example, with a weighting of $w_3 = 0.1$, corresponding to 10% priority given to execution time, and $w_1 = 0.9$, i.e. 90% priority given to vibrations, the optimal input parameters, located towards the bottom of the Pareto front, are as follows:

– cutting speed = 150 m/min;

– depth of cut = 0.5 mm;

– feed per revolution = 0.1 mm/rev.

Conversely, if the priorities are 90% given to execution time (w_3) and 10% to vibrations (w1), the new optimal parameters, located at the top of the Pareto front, are as follows:

– cutting speed = 140 m/min;

– depth of cut = 2 mm;

– feed per revolution = 0.4 mm/rev.

9.3. Conclusion

The optimization of industrial processes is a practice rooted in industrial culture for decades. Although it has evolved with the rise of digitalization, classical methods remain relevant and widely adopted.

The multi-objective optimization approach presented in this chapter results from the digitalization of industrial processes. Data collection from various sensors proves crucial for improving processes, having a significant impact on the entire production chain. Furthermore, cloud computing offers tools facilitating the design of algorithmic architectures, notably through the integration of artificial intelligence modules, whether predefined or specifically developed with common programming languages. In addition, it offers storage and computing capabilities indispensable for the success of this approach.

The results obtained demonstrate the full potential of cloud computing for optimizing digitalized industrial processes. In the present case, it allows production line supervisors to prioritize their optimization objectives according to various process parameters. Thanks to the flexible resources of cloud computing, it would be possible in the future to integrate other output criteria, such as machined surface quality, measured by roughness level. Indeed, the manufacturing parameters discussed in this chapter directly influence the state of machined surfaces. Depending on the importance given to final surface quality or to other additional criteria, supervisors will be able to develop models that allow optimization priorities to be adjusted to better meet quality requirements.

Although the implementation of this approach is complex, it proves the feasibility of such projects. However, it is essential to focus on the question of skills and training to ensure effective use of these advanced tools.

Conclusion

This work reveals that the digital transition is far from being a linear process. The experience acquired, through past and ongoing work, shows that submitting the same problem to different multidisciplinary teams can produce significantly varied results. Several factors explain this situation, notably a definition of expectations that is sometimes incomplete and subject to multiple interpretations.

For example, the digital twin is perceived by some as a simple immersive visualization tool without decision-making capacity, considering that decision-making processes fall under the Manufacturing Execution System (MES). Others, on the contrary, wish to endow the digital twin with decision-making modules, leaving the MES the role of executing instructions. Although these approaches are conceivable, they rely on fundamentally different functional architectures and communication systems, which generate uncertainties for companies.

The lack of universal standards and the desire of certain companies to dominate the market amplify these divergences. These companies, which offer cyber-physical industrial systems, intelligent Programmable Logic Controllers (PLCs), supervision solutions and digital platforms, seek to develop complete solutions. However, this can lead to technological dependency, often uncomfortable for companies wishing to evolve without losing the freedom to change suppliers.

This complexity underscores the need to structure the digital transition pathway of manufacturing companies in order to integrate the targeted functionalities while minimizing risks. This work presents cases of technological integration into existing production systems, with adapted methods and approaches. However, not all functionalities have been explored; only those considered priority have been the subject of in-depth studies.

Furthermore, the Industry 4.0 platform developed at the École Nationale Supérieure d'Arts et Métiers served as an experimental tool to test various technological, methodological and algorithmic approaches, in partnership with industrialists, institutions and technology providers. This work made it possible to overcome several concrete challenges, such as the need for complementary engineering studies generating additional costs, limitations in accessing data from certain machines or robots, as well as the complexity of interoperability between different systems. Recurrent malfunctions at the level of communication networks led to latencies, notably due to the difficult coexistence of different protocols. Other unforeseen events, such as lighting interfering with an optical scanner, were also encountered.

This work also offers pragmatic recommendations: avoid aiming for exhaustive integration of all functionalities; involve internal teams in the transition process, even though this may slightly slow down progress; anticipate system integration costs, often comparable to acquisition costs; and favor mature and sustainable technologies. It is also essential to budget for unforeseen but indispensable work.

In continuity with the presented work, several projects are underway and will enrich this system with new functionalities and technologies. Cybersecurity, for example, is becoming a major issue, requiring robust protocols and continuous monitoring to prevent intrusions. Furthermore, OT/IT convergence (operational technology and information technology) constitutes another central axis, aiming to harmonize physical and digital systems for more integrated and responsive management of industrial processes.

The training of engineers plays a key role in this transformation. Tomorrow's engineers will not only need to master traditional technical skills but also be trained in digital tools and advanced automation. However, it is difficult to achieve this level of versatility within a single training period. It therefore becomes essential to develop innovative, interactive and personalized training pathways, enabling future engineers to learn to interact with intelligent systems and to design coherent approaches integrating engineering, IT and cybersecurity.

Experiential learning through experimental platforms is also fundamental. These environments offer trainee engineers the possibility of deploying technological and methodological approaches in real conditions while collaborating with industrial actors. Engineers must also be capable of meeting the expectations of the different actors in the value chain, managing complex projects and designing solutions adapted to ethical and sustainability challenges. Finally, the ability to interact with various training platforms to enrich their skills remains a crucial challenge.

In summary, ongoing work is not limited to improving existing functionalities. It aims to create an interconnected, secure and agile industrial ecosystem, capable of responding to the current and future challenges of Industry 4.0. Training and the integration of new technologies will be determining factors for the success of this digital transformation and for the competitiveness of companies in a constantly evolving environment.

References

ABI Research (2024). 82 technology trends that will – and will not – shape 2024 [Online]. Available at: https://www.abiresearch.com/blogs/2024/01/30/current-and-future-iot-trends-and-statistics/ [Accessed 12 October 2024].

Aminharati, P. (2026). Physics-informed and data-driven modeling of composite hydrogen pressure vessels under extreme loading. PhD Thesis, École Nationale Supérieure d'Arts et Métiers, Paris.

Bencherki, Z. (2023). Méthodologie de conception et organisation de blocs fonctionnels en IEC 61499. PhD Thesis, École Nationale Supérieure d'Arts et Métiers, Paris.

Benfriha, K., El-Zant, C., Charrier, Q., Bouzid, A.-H., Wardle, P., Belaidi, I., Loubère, S., Ghodsian, N., Aoussat, A. (2021). Development of an advanced MES for the simulation and optimization of industry 4.0 process. *International Journal for Simulation and Multidisciplinary Design Optimization*, 12(23). doi: 10.1051/smdo/2021022.

Bertolini, M. and Guelfo, G. (2020). The Purdue Enterprise Reference Architecture: A framework for smart manufacturing. *Procedia CIRP*, 88, 329–334. doi: 10.1016/j.procir.2020.02.059.

Bonnard, R., Vieira, K., Arantes, M., Lorbieski, R., Nunes, M.C., Mattei, A.P. (2019). A big data/analytics platform for Industry 4.0 implementation in SMEs. EasyChair Preprint No. 1031 [Online]. Available at: https://easychair.org/publications/preprint/CKF1.

Bonnard, R., Arantes, M.D.S., Lorbieski, R., Vieira, K.M.M., Nunes, M.C. (2021). Big data/analytics platform for Industry 4.0 implementation in advanced manufacturing context. *International Journal of Advanced Manufacturing Technology*, 117(5), 1959–1973. doi: 10.1007/s00170-021-07834-5.

Charrier, Q. (2024). Conception et intégration du jumeau numérique. PhD Thesis, École Nationale Supérieure d'Arts et Métiers, Paris.

El-Helou, M. (2023). Comment le Cloud Computing peut accroître les performances d'un système de production 4.0 ? PhD Thesis, École Nationale Supérieure d'Arts et Métiers, Paris.

El Zant, C. (2021). Industrie 4.0 processus. Quel modèle pour une intégration réussie. PhD Thesis, École Nationale Supérieure d'Arts et Métiers, Paris.

Flyvbjerg, B., Budzier, A., Lee, J.S., Keil, M., Lunn, D., Bester, D.W. (2022). The empirical reality of IT project cost overruns: Discovering a power-law distribution. *Journal of Management Information Systems*, 39(3), 607–639. doi: 10.1080/07421222.2022.2064561.

Godsian, N. (2023). Development of a framework to integrate mobile manipulators as cyber-physical systems into existing production systems in the context of industry 4.0. PhD Thesis, École Nationale Supérieure d'Arts et Métiers, Paris.

Gorecky, D. and Weyer, S. (2016). System architecture for Industrie 4.0 production plants. White Paper, SmartFactory, Kaiserslautern.

Grand View Research (2023). Cloud computing industry data book [Online]. Available at: https://www.grandviewresearch.com/sector-report/cloud-computing-industry-data-book [Accessed 12 October 2024].

Grieves, M. (2005). Product lifecycle management: The new paradigm for enterprises. *International Journal of Product Development*, 2(1/2), 71–84. doi: 10.1504/IJPD.2005.007303.

Grieves, M. and Vickers, J. (2017). Digital twin: Mitigating unpredictable, undesirable emergent behavior in complex systems. In *Transdisciplinary Perspectives on Complex Systems*, Kahlen F.-J., Flumerfelt, S., Alves, A. (eds). Springer, Berlin, 85–113. doi: 10.1007/978-3-319-38756-7_4.

Hakam, N. (2024). Caractérisation de la maintenance prédictive. PhD Thesis, École Nationale Supérieure d'Arts et Métiers, Paris.

Halawi-Ghoson, N. (2025). A methodology for securing a manufacturing line based on risk assessment, in context of industry 4.0.

Heidel, R., Hoffmeister, M., Hankel, M., Döbrich, U. (2019). *Industrie 4.0: The Reference Architecture Model RAMI 4.0 and the Industrie 4.0 Component.* DIN Deutsches Institut für Normung e.V., Beuth Verlag GmbH, Berlin, Vienna, Zurich,.

Industry IoT Consortium (2022). The Industrial Internet Reference Architecture (Version 1.10). Industry IoT Consortium [Online]. Available at: https://www.iiconsortium.org/iira/.

Kritzinger, W., Karner, M., Traar, G., Henjes, J., Sihn, W. (2018). Digital twin in manufacturing: A categorical literature review and classification. *IFAC-PapersOnLine*, 51(11), 1016–1022. doi: 10.1016/j.ifacol.2018.08.474.

Lina, Y.J., Lana, C.B., Huanga, C.Y. (2019). A realization of cyber-physical manufacturing control system through industrial internet of things through industrial internet of things. In *25th International Conference on Production Research Manufacturing Innovation: Cyber Physical Manufacturing Cyber Physical Manufacturing*, Chicago.

Market.us. (2024). Global data lake market report [Online]. Available at: https://market.us/report/data-lake-market/.

MESA International (1997). MESA-11 model: A framework for manufacturing excellence [Online]. Available at: https://www.mesa.org.

MESA International (2016). MOM Best Practices Guide [Online]. Available at: https://www.mesa.org.

MESA International (2022). The New MESA Smart Manufacturing Model: Overview and Insights [Online]. Available at: https://www.mesa.org.

Nikooharf-Tamiz, M.H. (2024). Étude multi-échelles de l'optimisation intelligente de la stabilité dimensionnelle et des propriétés mécaniques de structures obtenues par fabrication additive. PhD Thesis, École Nationale Supérieure d'Arts et Métiers, Paris.

Purdue University (1992). Purdue Enterprise Reference Architecture. Report.

Shakeri, Z. (2025). Realistic multi-robot dynamic flexible job shop scheduling with finite intermediate buffers in context of Industry 4.0. PhD Thesis, École Nationale Supérieure d'Arts et Métiers, Paris.

Tao, F. (2018). Data-driven smart manufacturing. *Journal of Manufacturing Systems*, 48, 157–169. doi: 10.1016/j.jmsy.2018.01.006.

Terrazas, G., Ferry, N., Ratchev, S. (2019). A cloud-based framework for shop floor big data management and elastic computing analytics. *Computers in Industry*, 109, 204–214. doi: 10.1016/j.compind.2019.03.005.

Wang, X.V., Givehchi, M., Wang, L. (2017). Manufacturing system on the cloud: A case study on cloud-based process planning. *Procedia CIRP*, 63, 39–45. doi: 10.1016/j.procir.2017.03.103.

Westphall, C.B. (2021). AutoManSec 4 CloudIoT – Autonomic management and security for cloud and IoT. Research Project.

World Economic Forum (2019). Want to cut greenhouse gas emissions? Look to digital technologies [Online]. Available at: https://www.weforum.org/agenda/2019/09/want-to-exponentially-reduce-ghg-emissions-look-to-digital-tech-solutions/ [Accessed 12 October 2024].

World Economic Forum (2023). The future of jobs report [Online]. Available at: https://www.weforum.org/publications/the-future-of-jobs-report-2023/ [Accessed 12 October 2024].

Yong-Woon, K., Hyunjeong, L., Sangkeun, Y. (2021). Characterization of digital twin. In *IEEE International Conference on Industrial Engineering and Engineering Management*, 28–32. doi: 10.1109/IEEM50597.2021.9677997.

Index

A, B

C, D

P, Q

R, S, T

Other titles from

iSTE

in

Systems and Industrial Engineering – Robotics

2026

KRYSINSKI Tomasz, MALBURET François
Autonomous Vehicles: Aerospace and Automotive Applications

2025

MACHADO Carolina, DAVIM J. Paulo
From Industry 4.0 to Industry 6.0

2024

ANDRÉ Jean-Claude
Knowledge Production Modes between Science and Applications 1: Concepts
Knowledge Production Modes between Science and Applications 2: Applications

BRIFFAUT Jean-Pierre
Complexities 2: Various Approaches in the Field of Social and Human Sciences
(Systems of Systems Complexity Set – Volume 6)

OUSTALOUP Alain
Control in System Dynamics: Comparative Analysis of Feedback Strategies

YALAOUI Farouk, NGUYEN Nhan-Quy, OUAZENE Yassine, ZEMZAMI Maria, MERGHEM-BOULAHIA Leïla
Connected Innovation and Technology X.0 1: Advancing Digital Transformation: Innovational Strategies for Smart Industry and Sustainable Innovation

2023

BERRAH Lamia, CLIVILLÉ Vincent
Decision and Decision-maker in an Industrial Environment

BRIFFAUT Jean-Pierre
Complexities 1: Various Approaches in the Field of Techno-scientific Knowledge
(Systems of Systems Complexity Set – Volume 5)

CHAINEAUX Jacques
The Mechanisms of Explosions: 27 Case Studies for their Understanding

DIALLO Saikou, BOGUI Jean-Jacques Maomra, ATCHOUA Julien
Technologies and African Societies in Pandemic Times: Using Technology to Survive and Thrive During the Covid-19 Era
(Africa Digital Age Set – Volume 2)

JAEGLER Anicia, ROQUES Thierry
Sustainable Supply Chain Management: Sustainable Decision Support

PRINTZ Jacques
Organization and Pedagogy of Complexity: Systemic Case Studies and Prospects
(Systems of Systems Complexity Set – Volume 4)

2022

AMARA Yacine, BEN AHMED Hamid, GABSI Mohamed
Hybrid Excited Synchronous Machines: Topologies, Design and Analysis

BOURRIÈRES Jean-Paul, PINÈDE Nathalie, TRAORÉ Mamadou Kaba, ZACHAREWICZ Grégory
From Logistic Networks to Social Networks: Similarities, Specificities, Modeling, Evaluation

DEMOLY Frédéric, ANDRÉ Jean-Claude
4D Printing 1: Between Disruptive Research and Industrial Applications
4D Printing 2: Between Science and Technology

HAJJI Rafika, JARAR OULIDI Hassane
Building Information Modeling for a Smart and Sustainable Urban Space

KROB Daniel
Model-based Systems Architecting: Using CESAM to Architect Complex Systems
(Systems of Systems Complexity Set – Volume 3)

LOUIS Gilles
Dynamics of Aircraft Flight

2020

BRON Jean-Yves
System Requirements Engineering

KRYSINSKI TOMASZ, MALBURET FRANÇOIS
Energy and Motorization in the Automotive and Aeronautics Industries

PRINTZ Jacques
System Architecture and Complexity: Contribution of Systems of Systems to Systems Thinking

2019

ANDRÉ Jean-Claude
Industry 4.0: Paradoxes and Conflicts

BENSALAH Mounir, ELOUADI Abdelmajid, MHARZI Hassan
Railway Information Modeling RIM: The Track to Rail Modernization

BLUA Philippe, YALAOU Farouk, AMODEO Lionel, DE BLOCK Michaël, LAPLANCHE David
Hospital Logistics and e-Management: Digital Transition and Revolution

BRIFFAUT Jean-Pierre
From Complexity in the Natural Sciences to Complexity in Operations Management Systems
(Systems of Systems Complexity Set – Volume 1)

BUDINGER Marc, HAZYUK Ion, COÏC Clément
Multi-Physics Modeling of Technological Systems

FLAUS Jean-Marie
Cybersecurity of Industrial Systems

JAULIN Luc
Mobile Robotics – Second Edition Revised and Updated

KUMAR Kaushik, DAVIM Paulo J.
Optimization for Engineering Problems

TRIGEASSOU Jean-Claude, MAAMRI Nezha
Analysis, Modeling and Stability of Fractional Order Differential Systems 1: The Infinite State Approach
Analysis, Modeling and Stability of Fractional Order Differential Systems 2: The Infinite State Approach

VANDERHAEGEN Frédéric, MAAOUI Choubeila, SALLAK Mohamed, BERDJAG Denis
Automation Challenges of Socio-technical Systems

2018

BERRAH Lamia, CLIVILLÉ Vincent, FOULLOY Laurent
Industrial Objectives and Industrial Performance: Concepts and Fuzzy Handling

GONZALEZ-FELIU Jesus
Sustainable Urban Logistics: Planning and Evaluation

GROUS Ammar
Applied Mechanical Design

LEROY Alain
Production Availability and Reliability: Use in the Oil and Gas Industry

MARÉ Jean-Charles
Aerospace Actuators 3: European Commercial Aircraft and Tiltrotor Aircraft

MAXA Jean-Aimé, BEN MAHMOUD Mohamed Slim, LARRIEU Nicolas
Model-driven Development for Embedded Software: Application to Communications for Drone Swarm

MBIHI Jean
Analog Automation and Digital Feedback Control Techniques
Advanced Techniques and Technology of Computer-Aided Feedback Control

MORANA Joëlle
Logistics

SIMON Christophe, WEBER Philippe, SALLAK Mohamed
Data Uncertainty and Important Measures
(Systems Dependability Assessment Set – Volume 3)

TANIGUCHI Eiichi, THOMPSON Russell G.
City Logistics 1: New Opportunities and Challenges
City Logistics 2: Modeling and Planning Initiatives
City Logistics 3: Towards Sustainable and Liveable Cities

ZELM Martin, JAEKEL Frank-Walter, DOUMEINGTS Guy, WOLLSCHLAEGER Martin
Enterprise Interoperability: Smart Services and Business Impact of Enterprise Interoperability

2017

ANDRÉ Jean-Claude
From Additive Manufacturing to 3D/4D Printing 1: From Concepts to Achievements
From Additive Manufacturing to 3D/4D Printing 2: Current Techniques, Improvements and their Limitations
From Additive Manufacturing to 3D/4D Printing 3: Breakthrough Innovations: Programmable Material, 4D Printing and Bio-printing

ARCHIMÈDE Bernard, VALLESPIR Bruno
Enterprise Interoperability: INTEROP-PGSO Vision

CAMMAN Christelle, FIORE Claude, LIVOLSI Laurent, QUERRO Pascal
Supply Chain Management and Business Performance: The VASC Model

FEYEL Philippe
Robust Control, Optimization with Metaheuristics

MARÉ Jean-Charles
Aerospace Actuators 2: Signal-by-Wire and Power-by-Wire

POPESCU Dumitru, AMIRA Gharbi, STEFANOIU Dan, BORNE Pierre
Process Control Design for Industrial Applications

RÉVEILLAC Jean-Michel
Modeling and Simulation of Logistics Flows 1: Theory and Fundamentals
Modeling and Simulation of Logistics Flows 2: Dashboards, Traffic Planning and Management
Modeling and Simulation of Logistics Flows 3: Discrete and Continuous Flows in 2D/3D

2016

ANDRÉ Michel, SAMARAS Zissis
Energy and Environment
(Research for Innovative Transports Set – Volume 1)

AUBRY Jean-François, BRINZEI Nicolae, MAZOUNI Mohammed-Habib
Systems Dependability Assessment: Benefits of Petri Net Models
(Systems Dependability Assessment Set – Volume 1)

BLANQUART Corinne, CLAUSEN Uwe, JACOB Bernard
Towards Innovative Freight and Logistics
(Research for Innovative Transports Set – Volume 2)

COHEN Simon, YANNIS George
Traffic Management
(Research for Innovative Transports Set – Volume 3)

MARÉ Jean-Charles
Aerospace Actuators 1: Needs, Reliability and Hydraulic Power Solutions

REZG Nidhal, HAJEJ Zied, BOSCHIAN-CAMPANER Valerio
Production and Maintenance Optimization Problems: Logistic Constraints and Leasing Warranty Services

TORRENTI Jean-Michel, LA TORRE Francesca
Materials and Infrastructures 1
(Research for Innovative Transports Set – Volume 5A)
Materials and Infrastructures 2
(Research for Innovative Transports Set – Volume 5B)

WEBER Philippe, SIMON Christophe
Benefits of Bayesian Network Models
(Systems Dependability Assessment Set – Volume 2)

YANNIS George, COHEN Simon
Traffic Safety
(Research for Innovative Transports Set – Volume 4)

ZELM Martin, DOUMEINGHTS Guy, MENDONÇA Joao Pedro
Enterprise Interoperability in the Digitized and Networked Factory of the Future: I-ESA 16 Proceedings

2015

AUBRY Jean-François, BRINZEI Nicolae
Systems Dependability Assessment: Modeling with Graphs and Finite State Automata

BOULANGER Jean-Louis
CENELEC 50128 and IEC 62279 Standards

BRIFFAUT Jean-Pierre
E-Enabled Operations Management

CORSI Patrick, NEAU Erwan
Innovation Capability Maturity Mode

MISSIKOFF Michele, CANDUCCI Massimo, MAIDEN Neil
Enterprise Innovation

2014

BERGHEAU Jean-Michel
Thermomechanical Industrial Processes: Modeling and Numerical Simulation

CHETTO Maryline
Real-time Systems Scheduling
Volume 1 – Fundamentals
Volume 2 – Focuses

DAL PONT Jean-Pierre, AZZARO-PANTEL Catherine
New Approaches to the Process Industries: The Manufacturing Plant of the Future

DAVIM J. Paulo
Machinability of Advanced Materials

ESTAMPE Dominique
Supply Chain Performance and Evaluation Models

FAVRE Bernard
Introduction to Sustainable Transports

GAUTHIER Michaël, ANDREFF Nicolas, DOMBRE Etienne
Intracorporeal Robotics: From Milliscale to Nanoscale

LAURAS Matthieu, ZELM Martin, ARCHIMÈDE Bernard, BÉNABEN Frédérick, DOUMEINGTS Guy
Enterprise Interoperability: Interoperability for Agility, Resilience and Plasticity of Collaborations: I-ESA 14 Proceedings

MICOUIN Patrice
Model Based Systems Engineering: Fundamentals and Methods

MILLOT Patrick
Designing Human–Machine Cooperation Systems
Risk Management in Life Critical Systems

NI Zhenjiang, PACORET Céline, BENOSMAN Ryad, RÉGNIER Stéphane
Haptic Feedback Teleoperation of Optical Tweezers

OUSTALOUP Alain
Diversity and Non-integer Differentiation for System Dynamics

REZG Nidhal, DELLAGI Sofien, KHATAD Abdelhakim
Joint Optimization of Maintenance and Production Policies

STEFANOIU Dan, BORNE Pierre, POPESCU Dumitru, FILIP Florin Gh., EL KAMEL Abdelkader
Optimization in Engineering Sciences: Metaheuristics, Stochastic Methods and Decision Support

2013

ALAZARD Daniel
Reverse Engineering in Control Design

ARIOUI Hichem, NEHAOUA Lamri
Driving Simulation

CHADLI Mohammed, COPPIER Hervé
Command-control for Real-time Systems

DAAFOUZ Jamal, TARBOURIECH Sophie, SIGALOTTI Mario
Hybrid Systems with Constraints

FEYEL Philippe
Loop-shaping Robust Control

FLAUS Jean-Marie
Risk Analysis: Socio-technical and Industrial Systems

FRIBOURG Laurent, SOULAT Romain
Control of Switching Systems by Invariance Analysis: Application to Power Electronics

GROSSARD Mathieu, REGNIER Stéphane, CHAILLET Nicolas
Flexible Robotics: Applications to Multiscale Manipulations

GRUNN Emmanuel, PHAM Anh Tuan
Modeling of Complex Systems: Application to Aeronautical Dynamics

HABIB Maki K., DAVIM J. Paulo
Interdisciplinary Mechatronics: Engineering Science and Research Development

HAMMADI Slim, KSOURI Mekki
Multimodal Transport Systems

JARBOUI Bassem, SIARRY Patrick, TEGHEM Jacques
Metaheuristics for Production Scheduling

KIRILLOV Oleg N., PELINOVSKY Dmitry E.
Nonlinear Physical Systems

LE Vu Tuan Hieu, STOICA Cristina, ALAMO Teodoro,
CAMACHO Eduardo F., DUMUR Didier
Zonotopes: From Guaranteed State-estimation to Control

MACHADO Carolina, DAVIM J. Paulo
Management and Engineering Innovation

MORANA Joëlle
Sustainable Supply Chain Management

SANDOU Guillaume
Metaheuristic Optimization for the Design of Automatic Control Laws

STOICAN Florin, OLARU Sorin
Set-theoretic Fault Detection in Multisensor Systems

ZELM Martin, VAN SINDEREN Marten, PIRES Luis Ferraira, DOUMEINGTS Guy
Enterprise Interoperability: Research and Applications in the Service-oriented Ecosystem: I-ESA'13 Proceedings

2012

AÏT-KADI Daoud, CHOUINARD Marc, MARCOTTE Suzanne, RIOPEL Diane
Sustainable Reverse Logistics Network: Engineering and Management

BORNE Pierre, POPESCU Dumitru, FILIP Florin G., STEFANOIU Dan
Optimization in Engineering Sciences: Exact Methods

BOULANGER Jean-Louis
Industrial Use of Formal Methods

CHADLI Mohammed, BORNE Pierre
Multiple Models Approach in Automation: Takagi-Sugeno Fuzzy Systems

DAL PONT Jean-Pierre
Process Engineering and Industrial Management

DAVIM J. Paulo
Lasers in Manufacturing

DECLERCK Philippe
Discrete Event Systems in Dioid Algebra and Conventional Algebra

DOUMIATI Moustapha, CHARARA Ali, VICTORINO Alessandro, LECHNER Daniel
Vehicle Dynamics Estimation using Kalman Filtering: Experimental Validation

GUERRERO José A, LOZANO Rogelio
Flight Formation Control

HAMMADI Slim, KSOURI Mekki
Advanced Mobility and Transport Engineering

MAILLARD Pierre
Competitive Quality Strategies

MATTA Nada, VANDENBOOMGAERDE Yves, ARLAT Jean
Supervision and Safety of Complex Systems

POLER Raúl, CARNEIRO Luis Maia, JASINSKI Thomas, ZOLGHADRI Marc, PEDRAZZOLI Paolo
Intelligent Non-hierarchical Manufacturing Networks

TROCCAZ Jocelyne
Medical Robotics

YALAOUI Alice, CHEHADE Hicham, YALAOUI Farouk, AMODEO Lionel
Optimization of Logistics

ZELM Martin, SANCHIS Raquel, POLER Raúl, DOUMEINGTS Guy
Enterprise Interoperability: I-ESA'12 Proceedings

2011

CANTOT Pascal, LUZEAUX Dominique
Simulation and Modeling of Systems of Systems

DAVIM J. Paulo
Mechatronics
Wood Machining

GROUS Ammar
Applied Metrology for Manufacturing Engineering

KOLSKI Christophe
Human–Computer Interactions in Transport

LUZEAUX Dominique, RUAULT Jean-René, WIPPLER Jean-Luc
Complex Systems and Systems of Systems Engineering

ZELM Martin, VAN SINDEREN Marten, DOUMEINGTS Guy, JOHNSON Pontus
Enterprise Interoperability: IWEI2011 Proceedings

2010

BOTTA-GENOULAZ Valérie, CAMPAGNE Jean-Pierre, LLERENA Daniel, PELLEGRIN Claude
Supply Chain Performance: Collaboration, Alignment and Coordination

BOURLÈS Henri, GODFREY K.C. Kwan
Linear Systems

CHAILLET Nicolas, REGNIER Stéphane
Microrobotics for Micromanipulation

DAVIM J. Paulo
Sustainable Manufacturing

GIORDANO Max, MATHIEU Luc, VILLENEUVE François
Product Life-Cycle Management: Geometric Variations

LOZANO Rogelio
Unmanned Aerial Vehicles: Embedded Control

LUZEAUX Dominique, RUAULT Jean-René
Systems of Systems

VILLENEUVE François, MATHIEU Luc
Geometric Tolerancing of Products

WU Bing, BOURRIÈRES Jean-Paul
Proceedings of CEISIE'09

2009

DIAZ Michel
Petri Nets: Fundamental Models, Verification and Applications

ÖZEL Tugrul, DAVIM J. Paulo
Intelligent Machining

PITRAT Jacques
Artificial Beings

2008

ARTIGUES Christian, DEMASSEY Sophie, NÉRON Emmanuel
Resources–Constrained Project Scheduling

BILLAUT Jean-Charles, MOUKRIM Aziz, SANLAVILLE Eric
Flexibility and Robustness in Scheduling

DOCHAIN Denis
Bioprocess Control

JACOB Bernard, O'BRIEN Eugene, O'CONNOR Alan, BOUTELDJA Mohamed
HVParis 2008 - ICWIM5
HVParis 2008 - HVTT10

LOPEZ Pierre, ROUBELLAT François
Production Scheduling

THIERRY Caroline, THOMAS André, BEL Gérard
Supply Chain Simulation and Management

2007

DE LARMINAT Philippe
Analysis and Control of Linear Systems

DOMBRE Etienne, KHALIL Wisama
Robot Manipulators

LAMNABHI-LAGARRIGUE Françoise, LAGHROUCHE Salah, LORIA Antonio, PANTELEY Elena
Taming Heterogeneity and Complexity of Embedded Control

LIMNIOS Nikolaos
Fault Trees

2006

French College of Metrology
Metrology in Industry

Najim Kaddour
Control of Continuous Linear Systems

Printed and bound by CPI Group (UK) Ltd, Croydon, CR0 4YY
13/09/2026
14961013-0004